Poincaré's Philosophy

From Conventionalism to Phenomenology

ELIE ZAHAR

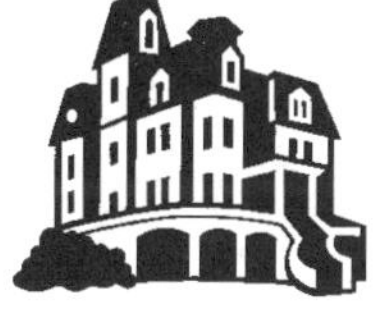

OPEN COURT
Chicago and La Salle, Illinois

To Helen, Isabelle, and George†,
who often spoke about this book

To order books from Open Court, call toll free 1-800-815-2280.

Open Court Publishing Company is a division of Carus Publishing Company.

First printing 2001

Printed and bound in the United States of America

Library of Congress Cataloging-in-Publication Data

Zahar, Elie.
Poincaré's philosophy : from conventionalism to phenomenology / Elie Zahar.
p. cm.
Includes bibliographical references and index.
ISBN 0-8126-9435-X (pbk. : alk. paper)
1. Poincaré, Henri, 1854–1912. I. Title.

B2430.P564 Z34 2001
194—dc21

2001021601

Brief Contents

Detailed Contents

Introduction

Speaking about Kant, Schopenhauer rightly claimed that it is far easier to criticize a genius than to praise him: his errors, always finite in number, can be easily circumscribed while his contribution to truth remains both inexhaustible and unfathomable (Schopenhauer 1958, Vol.1, Appendix). If I say this, it is by way of an excuse for the criticisms which will be levelled at some paradoxical aspects of Henri Poincaré's philosophy. However, my aim, which is a constructive one, is that of clarifying and of then reconciling Poincaré's various theses about the foundations of mathematics and of the natural sciences.

With Hilbert, Poincaré was the greatest mathematician of his time and one of the greatest ever. In E.T. Bell's words, he was the last universalist in that he looked upon every branch of mathematics and of theoretical physics as part of his domain (Bell 1937, Chapter 28). Poincaré had a prodigious knowledge of the field of complex analysis on which he left an indelible mark; he founded the discipline of *analysis situs* or *topology*; he studied the theory of groups, the probability calculus, classical mechanics, optics, electromagnetism, and astronomy; he attacked the three-body problem and addressed the question of planetary stability, thereby creating chaos theory. In this context, he established a result which proved essential to the second law of thermodynamics. Together with Lorentz and simultaneously with Einstein, he discovered Special Relativity. We owe him the Principle of Lorentz-covariance and one of the first applications of a symmetry requirement to physical laws; also the concept of an integrated four-dimensional physical continuum which was later attributed exclusively—and hence wrongly—to Minkowski. He determined the relativistic transformation equations for the electromagnetic field and for the charge-density, thereby correcting a serious mistake made by Lorentz in his 1904 paper. Poincaré finally developed a Lorentz-covariant gravitational theory which yielded Newton's law as a limiting case for small velocities. It is worth adding that he achieved these breakthroughs concurrently with, but independently of Einstein (see Chapter 4 below). In the history of French mathematics, Poincaré's phenomenal output was superseded only by Cauchy's.

Poincaré wrote several articles and books on epistemology and on the psychology of discovery. Like Einstein, he offers the rare example of a creative scientist whose work is informed by an explicit methodological standpoint. Poincaré was described—by Popper—as the greatest philosopher of science ever (see Bouveresse 1998, p. 191). He was however no systematic philosopher, or rather: he did not have enough time to become one. As a result, his philosophical pieces, written at different times and in different circumstances, contain several incoherencies while his superbly lucid style conceals the complexity and depth of his thought. These apparent paradoxes prevented Poincaré's contemporaries from appreciating his philosophical insights at their true value. His theses nonetheless helped to shape the philosophy which was to guide the development of science throughout the twentieth century; for it can be maintained that the Neopositivists—the members of the Vienna and Berlin Circles—did no more than 'precisify' Poincaré's (and Duhem's) insights into the problems of induction, of the empirical basis, and of the mathematical foundations of physics.

Going back to the purpose of this book—a purpose which might make its reading worthwhile to philosophers, to historians and even to working scientists—it consists of an exposé followed by a rational reconstruction of Poincaré's position. The book can also be regarded, from a non-exegetic standpoint, as a treatise on modern epistemology.

The following passages, both taken from the Introduction to *The Value of Science,* demonstrate the necessity of reinterpreting Poincaré's philosophy; for these quotations, which expose his conflicting epistemological views in their most radical form, illustrate the 'Problematik' underlying the whole of the present work:

> The search for truth should be the goal of our activities. It is the sole end worthy of them. . . .
>
> Some people have exaggerated the role of convention in science, they have gone so far as to say that law, that scientific fact itself, was created by the scientist. This is going much too far in the direction of nominalism. No, scientific laws are not artificial creations; we have no reason to regard them as accidental, though it be impossible to prove they are not . . .
>
> Does the harmony the human intelligence thinks it discovers in nature exist outside of this intelligence? No, beyond doubt, a reality completely independent of the mind which conceives it, sees or feels it, is an impossibility. (Poincaré 1906a, pp. 11, 14)

Let me now give a few details about the contents of the present book. I propose first to examine two areas in which Poincaré's philosophy of physics came into close contact with his theses about the foundations of mathematics. By not strictly distinguishing between physical space and the mathematical continuum, he was led to a neo-Kantian conception of geometry; and because of his reliance on Kant's views concerning our intuition of time, he assessed the status of certain mathematical postulates, more particularly that of mathematical induction, in a manner which remains highly questionable (see Chapters 3, 5 below).

Poincaré took the mathematical and the empirical logics of discovery to be essentially the same. He stressed the universal role of intuition by which we *guess* the solution to a problem *before* trying either to establish or to refute our conjecture. Prima facie, this approach presupposes a belief in the mind-independent existence *both* of a physical *and* of a mathematical reality; but despite his structural-realist stance in the philosophy of physics, Poincaré insisted not only that *mathematical* entities must in principle be constructible, but also that they ought actually to have been constructed before we can meaningfully talk about them (see Chapters 2, 5 below). In Kantian fashion, he held that infinite classes are not completed entities about which we are entitled to reason; for they remain open-ended collections to which new members are continually added by discrete acts of consciousness. Thus Poincaré defines *all* aggregates in strictly *temporal* terms: we cannot legitimately refer to any infinite classes, for these are continually altered by the introduction of new elements; and this constant change takes place because no *mathematical* entity exists prior to its construction in pure time. Mathematical objects are thus thrown into existence by the activity of the human mind; they do not antedate its operations. If we want mathematical statements to retain some objective validity, Poincaré seems to leave us no choice but to postulate a set of universal rules to which everybody's mind conforms; a thesis which was in fact refuted by Poincaré's own controversies with both Russell and Zermelo.

Poincaré thus held a quasi-Kantian view of mathematics. He must nonetheless have known that in classical analysis, a field to which he made fundamental contributions, the real numbers are defined by means of *infinite* sets of fractions or of *infinite* classes of Cauchy sequences. His practice was therefore in open conflict with his philosophical position; and his constructivism appears philosophically more dubious than Platonist realism which postulates a realm of mind-independent mathematical entities (see Chapter 6 below). From Poincaré's own viewpoint, Platonism would moreover have had the merit of treating the logics of scientific and of mathematical discovery on a par: in both cases, the mind would be trying to capture a reality existing prior to its operations. Moreover, only Platonism seems capable of making disagreements between mathematicians a matter for rational discussion rather than for a mere recognition of irreducible differences between personal preferences. Needless to say, there remains one important contrast between physics and mathematics: scientific hypotheses can be subjected to empirical tests and hence to the verdict of a jury over whose decisions we have no complete control. Such tests are unavailable to mathematicians; apart from a minimal condition of consistency, the intuitive self-evidence of the theorems must therefore continue to play a central role in the progress of mathematics.

Another object of this book is to isolate some seemingly incompatible strands in Poincaré's philosophy of physics. I shall distinguish between his so-called conventionalism on the one hand, and a marked tendency towards an inductivist form of empiricism on the other. It can be shown that most of these conflicts can be resolved, that Poincaré was a structural realist with strong empiricist leanings, and that the more carefully one analyses his conventionalism, the more it assumes the form of a *'pétition de principe'*. In certain contexts, Poincaré held that high-level theories are neither true nor false but more or less convenient *and*—in a different context—that these selfsame hypotheses prove convenient only to the extent that they mirror reality. Thus he treated the degree of convenience of physical principles, their level of aesthetic-cum-mathematical perfection, as an index of their verisimilitude. (It should be immediately added that he treated 'verisimilitude' as an intuitive concept unrelated to the correspondence theory of truth). Though his conventionalism seemed bound to lead to epistemological relativism, Poincaré remained a structural realist through and through (see Chapter 2). He insisted on rejecting *as false* all experimentally undermined laws. As for his principle of convenience *(commodité)*, it enjoined him to accept unified scientific systems only as long as they remained strongly supported by *crude* facts (see Chapter 1).

Poincaré's views will often be compared and contrasted with those of Duhem (see Chapter 2). This is why it seems appropriate that I should end this introduction by saying something about the different meanings given by these two influential authors to the word 'conventionalism'. In a nutshell, Duhem's thesis is that the human mind is intrinsically incapable of grasping reality as she really is, as created by God; for the divine mind transcends all human understanding. The most that a scientist can hope to achieve is some *natural classification* of the phenomena, a scheme which simulates the ontological order without actually referring to it. According to

Duhem, knowledge is definitely not an explanation but a coherent pigeon-holing of appearances. We shall see that this attitude contrasts with Poincaré's position—as rationally reconstructed. Unlike Duhem who tried to make the world safe for a specific theology, Poincaré had *a priori* no *metaphysical* axe to grind; but right from the start, one important aim of his philosophy was to account for the creativity of man's scientific theorizing. In this respect, he can again be compared to Kant who tackled the problem posed by the presumed infallibility of certain components of human knowledge. Kant's well-known answer was that man's *spontaneous* or *creative* mind prescribes its own laws to physical nature. Poincaré was similarly concerned with the—admittedly relative—autonomy of thought vis-à-vis the world of sense-experience. In the last analysis, his Conventionalism merely serves to emphasize the mind's active and largely independent role in scientific discovery. 'Conventionalism' actually turns out to be a misnomer which ought to be replaced by something like a 'Principle of Mathematical Coherence'.

1

Poincaré's General Philosophy of Science

It is well-known that Poincaré's first and most frequently cited philosophical work bears the title *Science and Hypothesis* (Poincaré 1902). It is therefore appropriate that we should start by giving a largely nontechnical account of Poincaré's general views about the role of theories and of empirical laws in the physical sciences.

1.1 The Character of Physical Theory

For Poincaré as for Popper, it is impossible to pursue science without preconceived ideas, without provisional conjectures:

> It is often said that experiments should be made without preconceived ideas. This is impossible. Not only would it make every experiment fruitless, but even if we wished to do so, it could not be done. Every man has his own conception of the world, and this he cannot so easily lay aside. We must, for example, use language, and our language is necessarily steeped in preconceived ideas, which are a thousand times the most dangerous of all. (Poincaré 1902, p. 143)

Poincaré divides scientific hypotheses into three categories. The first consists of a number of postulates "which are quite natural and necessary. It is difficult not to suppose that the influence of very distant bodies is quite negligible, that small movements obey a linear law and that effect is a continuous function of its cause" (Poincaré 1902, p. 152). Among these assumptions—which should be the last to be given up—figure certain conventions such as the axioms of Euclidean geometry and those of rational mechanics.

The second category comprises hypotheses described as indifferent because they perform a purely psychological function. Such are a number of metaphysical theses which assist our understanding and simplify our calculations, but make a difference neither to the mathematical form of theories nor to their empirical consequences.

> For example, in most questions the analyst assumes, at the beginning of his calculations, either that matter is continuous, or the reverse, that it is formed of atoms. On the atomic supposition, he has a little more difficulty in obtaining them—that is all. (Poincaré 1902, p. 152)

The third and last category consists of empirically refutable generalisations which are inductively obtained from scientific facts.

1.2 Unity versus Induction

According to Poincaré, the very possibility of physical science rests on the twin principles of the unity and of the uniformity of nature. The first principle, which presupposes the interdependence of all physical processes, is absolutely indispensable. It constitutes a kind of Kantian a priori truth; whence the methodological primacy of the unity requirement over all others:

> Let us first observe that every generalisation supposes in a certain measure a belief in the unity and simplicity of Nature. As far as the unity is concerned, there can be no difficulty. If the different parts of the universe were not as the organs of the same body, they would not react one upon the other; they would mutually ignore each other, and we in particular should know only one part. We need not therefore ask if Nature is one, but how it is one. (Poincaré 1902, p. 145)

The following passage shows that Poincaré holds the degree of unity of a given hypothesis to be very different from that of its simplicity understood in its pragmatic sense:

> With this tendency there is no doubt a loss of simplicity. Such and such an effect was represented by straight lines; it is now necessary to connect these lines by more or less complicated curves. On the other hand, unity is gained. Separate categories quieted but did not satisfy the mind. (Poincaré 1902, p. 182)

What constitutes the unity of a theory is a sort of organic compactness arising from the interconnectedness of its parts. From a pragmatic viewpoint, such a theory is not necessarily simpler than its rivals. Let us now examine the principle of induction which entitles us to proceed from the particular to the general, thus enriching the content of some of our empirical premises. Like many classical empiricists, Poincaré believed that without induction, science could not have got off the ground:

> All laws are therefore deduced from experiment . . . (Poincaré 1906a, p. 76)

This principle plays an important role in the selection of facts, which scientists do not choose at random but in terms of the generalisations they might give rise to. This is why careful observers home in on simple facts having a great chance of recurring. For example: biologists study cells in preference to whole organisms because the former present simple patterns which lend themselves to easy generalisations. Any—necessarily finite—sequence of experimental results can however be generalised in infinitely many ways; which is why we need some rule for restricting the range of possibilities. The uniformity principle is one such rule which enjoins us to connect a set of points by means not of an arbitrary line but of a differentiable curve.

Understood in this way, induction is an indispensable tool for the researcher; and it might be thought that Poincaré looked upon it as a heuristic device having no metaphysical import. This is however far from being the case, for he repeatedly maintained that science aims at the truth. Hence a physicist who pursues such an aim while generalising from a finite sample must, if only implicitly, believe in the uniformity of nature. Poincaré then asks the question, meaningful only to a realist, whether the simplicity of the physical world is profound or only apparent. He notes that the two principles of unity and of uniformity often pull in opposite directions: if nature is one, then since any physical process is connected to every other entity in the universe, it must depend on a large number of parameters and thus obey very complex laws. Given the primacy of the unity principle, we have to ask ourselves whether we can proceed *as if* nature were simple. The second law of thermodynamics assures us that we can do so at no great risk; for according to Poincaré, nature tends towards more *apparent* uniformity arising from the intimate blending of many complex factors.

> The laws which govern visible bodies would then be mere consequences of the molecular laws. Their simplicity would then be merely apparent and would conceal an extremely complex reality since its complexity would be measured by the very number of molecules. But it is precisely because this number is very large that the divergences in details would be mutually compensatory and we would therefore believe that there was harmony. (Poincaré 1913a, p. 10)

1.3 The Falsifiability of Scientific Laws

We shall now examine the crucial problem posed by the status of empirical laws. Once again, Poincaré's attitude proves highly ambiguous. In certain passages, he holds generalisations to be both verifiable and refutable; in others, that for logical reasons experience can at most *disconfirm* experimental laws. Let us start with the problem of the refutability of scientific laws which is relatively straightforward. Poincaré's falsificationism turns out to be as intransigent—not to say as extreme—as Popper's. The following two passages could easily have been written by the author of *The Logic of Scientific Discovery:*

> Thus when a rule has been established, we have first to look for the cases in which the rule stands the best chance of being found at fault. This is one of many reasons for the interest of astronomical facts and of geological ages. By making long excursions in space and time, we may find our ordinary rules completely upset. . . . (Poincaré 1908, pp. 20–21)

And:

> Ah well! this impatience is not justified. The physicist who has just given up one of his hypotheses should, on the contrary, rejoice, for he has found an unexpected opportunity of discovery. (Poincaré 1902, p. 150)

Note that according to Poincaré, the descriptive terms occurring in empirical laws must possess meanings fixed independently of actual experimental results. He would have contemptuously rejected the artifice consisting, say, in altering the meaning of the word 'swan' *after* the proposition 'All swans are white' had been refuted. He had in effect put a ban on *the immunising stratagems* which were later denounced by Karl Popper and Hans Albert (Popper 1959, pp. 81–84; also Albert 1975, pp. 29–36). Arguing against E. Le Roy, Poincaré wrote:

> Recall first the examples he [Le Roy] has given. When I say: Phosphorus melts at 44°, I think I am enunciating a law; in reality, it is just the definition of phosphorus; if one should discover a body which, possessing otherwise all the properties of phosphorus, did not melt at 44°, we should give it another name, that is all, and the law would remain valid. . . . It is clear that, if laws were reduced to that, they could not serve in prediction; then they would be good for nothing, either as means of knowledge or as principles of action. (Poincaré 1906a, pp. 122–23)

And:

> There is no escape from this dilemma: either science does not enable us to foresee, and then it is valueless as a rule of action; or else it enables us to foresee in a fashion more or less imperfect, and then it is not without value as a means of knowledge. (Poincaré 1906a, p. 115)

Let us go back to the question of the alleged verifiability of empirical laws. In his 1902, Poincaré claimed that the "hypotheses of the third category are real generalisations. They must be confirmed or invalidated by experience. Whether verified or condemned, they will always be fruitful . . ." (Poincaré 1902, p. 153). But both in his 1906a and in his 1913a, he realised that, despite being refutable, low-level hypotheses can never be definitively established by experiment. Commenting on the possibility of laws changing over time, he moreover wrote:

> Thus there is not a single law which we can enunciate with the certainty that it has always been true in the past with the same approximation as today; in fact not even with the certainty that we shall never be able to demonstrate that it has been false in the past. And yet, there is nothing in this to prevent the scientist from maintaining his belief in the principle of immutability, since no law will ever be relegated to the rank of being transitory without being replaced by another law more general and more comprehensive. (Poincaré 1913a, p. 13)

Once again, Poincaré's position must be rectified in order to become coherent. Let us retain the view that the primitive terms occurring in empirical laws must have unalterable meanings, that is, denotations fixed in advance of experiment. Poincaré admits that these laws are—*by right*—either true or false; but since generalisations are universal propositions, experience might refute but certainly never verifies them. With regard to empirical laws, Poincaré did not therefore adopt a verificationist definition of meaning but instinctively adhered to the fully-fledged correspondence theory of truth.

1.4 Conventionalism and the Empiricist Meaning-Criterion

In order to understand the role and status which Poincaré attributes to scientific conventions, it proves useful to examine his critique of classical mechanics. In *Science and Hypothesis* he wonders whether Newtonian dynamics constitutes a rational discipline or an experimental science. He first notes that in our world there exist no strictly isolated bodies; that is: no systems entirely removed from the action of all external forces. It follows that the Principle of Inertia cannot be confirmed without recourse to some auxiliary assumptions. Poincaré concludes that since this principle is not independently testable, it does not constitute a proposition possessing a definite truth-value.

Consider next the law of motion: $\underline{f} = m\underline{a}$. Even if this equation were to be treated as a definition of force, we would still have to determine the scalar m, in other words, to provide some method of *measuring* the mass of the mobile. Poincaré resorts to Mach's definition according to which, A and B being any two bodies isolated from the rest of the universe, we set $m_A/m_B = -\underline{a}_B/\underline{a}_A$; where m_A and m_B, $\underline{a}_A$ and $\underline{a}_B$ denote the masses and accelerations of A and B respectively. Poincaré remarks that such a definition brings into play Newton's third law according to which every action provokes an equal and opposite reaction: we must have $m_A \cdot \underline{a}_A = -m_B \cdot \underline{a}_B$, where this equation is taken to hold independently of the initial positions and velocities of A and B. The Principle of Reaction must therefore be presupposed. In Poincaré's words, it will be treated as a *definition.* Given that A and B cannot be shielded against all external influences, account should moreover be taken of the exogenous factors which might affect the state of the two mobiles. In other words: one ought to consider not only the interactions between A and B but also the forces acting on the dynamical system from without. One would thus be relying on Newton's fourth law, according to which any effect caused by a third body C does not alter the interaction between A and B but is linearly added to it. This law must consequently be used as yet another *definition.*

From the above considerations, Poincaré concludes that the principles of mechanics are conventions and hence have nothing to fear from any new experiments we might be led to carry out. In *The Value of Science*, he retreats from this untenable position by conceding that Newton's laws might *globally* be invalidated by experience; and he takes this opportunity in order to define the role of conventions in the sciences:

> Certain laws have rendered so many services in the past that we elevate them to the rank of principles: in other words, we turn them into definitions; we no longer demand that they should hold on their own or else be falsified, but they enter as components into more comprehensive theories. Suppose the astronomers discover thart the stars do not exactly obey Newton's law. We can break up this proposition

> (1) 'The stars obey Newton's law' into two others: (2), 'gravitation obeys Newton's law', (3). 'gravitation is the only force acting on the stars'. In this case, proposition (2) is no longer anything but a definition and is beyond the test of experiment; but then it will be on proposition (3) that this check can be exercised. This is indeed necessary since the resulting proposition (4) predicts verifiable facts in the rough. (Poincaré 1906a, p. 124)

This passage contains an implicit reference to the Duhem-Quine problem, that is, to a dilemma which flows from an elementary logical rule: whenever a conjunction of propositions is undermined, we know that the whole conjunction is false but not which of the conjuncts is—or are—to blame; we can therefore decide to direct 'the arrow of the Modus Tollens' away from any chosen conjunct, which is thereby marked out as a *convention*. Poincaré however adds a word of caution: the decision to change a law into a definition or into a convention is legitimate only if it proves convenient. To repeat: he strongly opposed all stratagems designed to put a theory beyond the reach of experimental refutation. He recognized that such adhoc devices, though nearly always possible, destroy the predictive power of a theory by preventing it from forbidding any state of affairs. Commenting on such adhoc manoeuvres with regard to the law of the conservation of energy, he wrote:

> But then, what have we gained by this stroke? The principle is intact, but henceforth what use is it? It enabled us to foresee that in such circumstance we could count on such a total quantity of energy; it limited us; but now . . . we are no longer limited by anything; . . . if a principle ceases to be fecund, experiment, without contradicting it directly, will nevertheless have condemned it. (Poincaré 1906a, p. 110)

We now face the problem of explaining why, despite acknowledging that a global system can be undermined, Poincaré was reluctant to ascribe any truth-value to it and to some of its components. His position can be rationally reconstructed by attributing to him the following empiricist meaning criterion, which was implicitly held rather than explicitly stated.

Let us call 'simple proposition' any statement possessing no proper sentential components. Such a formula, which either is atomic or else starts with a quantifier, will be regarded as possessing a truth-value if it is verifiable or falsifiable or both. In other words: a simple proposition will be held to be *true-or false* iff it is at least partially decidable. As for molecular statements, they will be taken to have a truth-value iff all their sentential components have the same property. Note that this criterion is language and even formulation-dependent. Let us for instance write $\vdash p \Leftrightarrow [(q \vee p) \wedge (\neg q \vee p)]$, where q is totally undecidable while p is only refutable; the left, but not the right-hand side can therefore be a truth-bearer; which is a somewhat paradoxical result. One has nonetheless the feeling that this kind of criticism would not have overly worried Poincaré: for he looked upon the syntactical formulation of a theory as contributing to its content, or rather: to its capacity for simulating the real order of things. Concerning the above logical equivalence, he would have held the right-hand side to be not only cumbersome *but also nonsensical.* It would however be a mistake to assume that Poincaré adhered to the verificationist principle—as subsequently proposed by the Vienna Circle—which identifies the meaning of a sentence with the method of its verification. Far from it; for since most scientific propositions are universal and hence unverifiable, it could be maintained that Poincaré came very near to adopting a falsificationist definition of meaning. Such a definition should not be confused with Popper's demarcation criterion according to which a statement is scientific iff it has empirically decidable consequences. Popper held every syntactically well-formed formula to be interpretable and hence susceptible of truth or falsity. Since science and *good* metaphysics are both meaningful, Popper's falsificationist demarcation criterion delineates their common frontier. Not so for Poincaré. For him, descriptions of crude facts are fully decidable and hence pose no problem. Though unverifiable, experimental laws can be empirically falsified and are therefore—by right—either true or false. As for purely existential statements, which are verifiable but irrefutable, Poincaré hardly concerns himself with them, no doubt because they are largely irrelevant to theoretical science. Note that the conventions which play a central role in physics can be neither verified nor refuted and hence should, according to Poincaré, be deprived of any truth-value.

As is well-known, Tarski's truth-definition consists first in fixing a domain of individuals; then in analysing a proposition into primitive symbols each of which receives—separately—some prescribed meaning; finally, in assigning a truth-value to the recomposed proposition. For Poincaré, such an atomising process ceases to have any legitimacy as soon as one deals with totally undecidable statements. The latter can however act as metaphors which should be judged adequate if they prove convenient. By means of its syntactical structure, a good convention mirrors a transcendent reality which it cannot directly signify. Convenience, and convenience alone, should be treated as an index of verisimilitude; in other words, a hypothesis proves to be convenient to the extent that it comes closer to the truth. Let us nonetheless repeat that Poincaré makes use of an intuitive concept of verisimilitude which is independent of the correspondence theory; it is a syntactical notion based on the linguistic structure of a sentence and not on any semantic relation between the latter's descriptive terms and some external reality (see below, Chapter 2). Thus Poincaré subscribes to a *quasi*-Kantian thesis which allows us to simulate—though not directly to denote—the noumenal world. To those who might object that he uses unacceptably woolly concepts, Poincaré might well have replied that this is the best one can do and that these intuitive notions are anyway unavoidable in the sciences. Whatever the case may be, this form of Kantianism has considerable *methodological* implications. Since conventions do not strictly refer to *anything* at all, they are *a fortiori* incapable of describing the essence of things as they really are. This is why Poincaré was a throughgoing anti-essentialist. He thought it illegitimate to adduce, in favour of a hypothesis, that it posits a particularly intelligible ontology; for no convention should, at any rate *initially*, have any metaphysical pretensions: only its degree of unity and its predictive power reveal, *post-hoc,* its *truth-likeness.* Poincaré was therefore ruthless in rejecting a theory as soon as it proved inconvenient, and this no matter how 'intelligible' it might initially have appeared to be; for he attributed an exclusively psychological significance to the notion of intelligibility. As regards the relationship between verisimilitude on the one hand, and convenience and simplicity on the other, he maintained that:

> The imposing simplicity of Mayer's principle equally contributes to strengthen our faith. In a law immediately deduced from experiment, such as Mariotte's law, this simplicity would rather appear to us a reason for distrust; but here this is no longer the case. We take elements which at first glance are unconnected; these arrange themselves in an unexpected order and form a harmonious whole. We cannot believe that this unexpected harmony is a mere result of chance. (Poincaré 1902, p. 131)

As compared and contrasted with Lorentz, Poincaré had no deep commitment to the ontology of classical physics. For him, the laws of dynamics were conventions and could therefore not be strictly true of any ontology. Despite having initially claimed that classical mechanics will always prove to be the most convenient system, he realized that *absolute time* in no way constitutes a necessary a priori concept; for we cannot directly intuit the simultaneity of two events which either are spatially separated or else take place in two distinct minds. Only the temporal order within one consciousness is pregiven in the sense of not depending on our volition. But the notion of physical time requires a stipulation; and Poincaré's acceptance of Euclidean Geometry and of the absolute time *convention* was only provisional (see Poincaré 1902, p. 91). Although the

Galilean Transformation was regarded by Lorentz as the only 'true' one, Poincaré did not hesitate in throwing it overboard: as soon as it proved inconvenient, he replaced it by the Lorentz Group which alone was operative (see below, Chapter 4).

1.5 Convenience, Non-Adhoc Explanations, and the Principle of Unity

Let us try to clarify the notion of convenience which plays a central role in Poincaré's philosophy of science. This general concept possesses two distinct aspects. At the theoretical level, the degree of convenience of a hypothesis is identical with that of its unity; it reveals itself through the multiple connections linking the various components of a scientific system (see Appendix 3). It has already been noted that theoretical unity differs from pragmatic simplicity. At the empirical level, a theory will be judged convenient if it makes novel predictions or else establishes unexpected connections between known facts; all this, without resorting to auxiliary assumptions which are at odds with the central hypothesis. We have here two complementary aspects of convenience: a theory is convenient to the extent that it accounts for *experience* while preserving its *unity*. Under the impact of facts, a system might eventually disintegrate; in which case one should avoid local manoeuvres whose net effect is to fragment a theory through artificially bringing it into line with known empirical results. It is far preferable to build anew while abiding by the Correspondence Principle, which Poincaré defines and justifies as follows: should an old hypothesis H prove to have been systematically 'convenient' throughout some domain Δ, then it is improbable that this should be due to mere chance. H must be taken to express true relations which ought to reappear—perhaps in a slightly modified form—in some new theory T. In other words: T must tend to H whenever certain parameters tend to 0, thereby confining T to the domain Δ (see below, Chapter 4). Realising that classical thermodynamics is probably false and would therefore have to be replaced by an atomic hypothesis, Poincaré wrote:

> These conceptions up to now have always been confirmed by experiment and the proofs today are numerous enough that they cannot be attributed to chance. It will therefore be necessary, if new experiments reveal exceptions, not to abandon the theory but to modify it, to make it more comprehensive so as to permit it to include new facts. (Poincaré 1913a, p. 77)

The following quotation, taken from a paper presented at the 1904 Saint Louis Conference, summarises Poincaré's conception of scientific revolutions and his views concerning the role of conventions in mathematical physics:

> Suppose now that all these efforts fail and, after all, I do not believe that they will, what must be done? Will it be necessary to seek to mend the broken principles by giving what we French call a coup de pouce? That evidently is always possible and I retract nothing of what I said above . . . This, therefore, is not what would have to be done; it would be necessary to rebuild anew . . . As I said, we have already passed through a similar crisis. I have shown you that in the second mathematical physics, that of the principles, we find traces of the first, that of central forces; it will be just the same if we must know a third. Just so with the animal that exuviates, that breaks its too narrow carapace and makes itself a fresh one, under the new envelope one will recognize the essential traits of the organism which have persisted. (Poincaré 1906a, pp. 109–110)

In brief: scientific progress does not consist in carrying out modifications designed to rescue an existing theory but in starting afresh, while making sure that the old and empirically successful laws are limiting cases of the new ones.

1.6 Adhocness and Parameter-adjustment

We have seen that Poincaré declares a theory non-adhoc if it makes unexpected empirical predictions. Let us examine whether 'unexpectedness' can be given a non-psychologistic interpretation. In what follows, we shall try to explicate rigorously what it means for a hypothesis to be empirically adhoc; that is: to be obtained by the *post-hoc* adjustment of free parameters to known experimental results.

Consider a hypothesis $H(a_1,...,a_n)$ containing the parameters $a_1,...,a_n$; that is: $a_1,...,a_n$ are unknown prior to experimentation; and let $e_1,..., e_m$ be a sequence of empirical results. Since $a_1,...,a_n$ have not yet been fixed, we might initially be ignorant as to whether or not $H(a_1,...,a_n)$ subsumes e_1, e_2, ... and e_m. Through putting $a = (a_1,...,a_n)$ and $E \equiv e_1 \wedge ... \wedge e_m$, we can, without any loss of generality, assume that $n = m = 1$. We now face the task of so determining a that: (1) H(a) is consistent and: (2) $\vdash H(a) \Rightarrow E$.

Since scientists take E for granted while provisionally accepting H(a) as true, they will postulate the conjunction $H(a) \wedge E$, from which they then draw conclusions as to the possible values of a. They typically succeed in determining a set K such that: $\vdash (H(a) \wedge E) \Rightarrow (a \in K)$. I.e. $\vdash (a \notin K) \Rightarrow (H(a) \Rightarrow \neg E)$. Unless $a \in K$, H(a) will thus be refuted by the facts E. But scientists want E to be not merely consistent with, but also explained by H(a). So they will construct another set Q such that: $\vdash (a \in Q) \Rightarrow (H(a) \Rightarrow E)$. The next step usually consists in selecting some a_0 which satisfies $\vdash a_0 \in K \cap Q$. Since $\vdash a_0 \in Q$, we have: $\vdash H(a_0) \Rightarrow E$; i.e. $H(a_0)$ entails E. As for $a_0 \in K$, it represents a consistency condition. More precisely: if $\vdash a_0 \in Q$ but $\vdash a_0 \notin K$, then by the above: $\vdash H(a_0) \Rightarrow E$ and $\vdash H(a_0) \Rightarrow \neg E$; which means that $H(a_0)$ is logically or mathematically inconsistent. Thus, choosing $a_0 \in K \cap Q$ comes down to constructing a (hopefully) consistent theory $H(a_0)$ which is specifically doctored to subsume the facts E.

In such a situation, Poincaré would deny that H derives any support from E. Popper shared Poincaré's appraisal for which he offered the following rationale. Since E was known prior to $H(a_0)$, it formed part of the background knowledge B, available before $H(a_0)$ was put forward. Hence $p(E,B) = 1$. A fortiori: $p(E,H(a_0) \wedge B) = 1$. Hence: $C(H(a_0),E,B) =_{Def.} (p(E,H(a_0) \wedge B) - p(E,B)) / (p(E,H(a_0) \wedge B) + p(E,B) - p(H(a_0) \wedge E,B)) = 0$; where $C(H(a_0),E,B)$ is Popper's expression for the degree of the corroboration of $H(a_0)$ by E, given background knowledge B (see Popper 1959, p. 409).

Despite Poincaré's and Popper's sure methodological instincts, I think that the above argument must be rejected. Let us, for example, take E to be a description of Michelson's result, of the Balmer series, or of the precession of Mercury's perihelion. E would therefore have been known prior to Special Relativity, to Bohr's quantum theory and to General Relativity respectively (see Zahar 1989, pp. 13–17, 87–92). Thus in each of the three cases, E belongs to the extant background knowledge and Popper's degree of corroboration must consequently vanish. Yet we have the strong intuition that each of the three E's confirms the corresponding H; where the *rationale* behind this intuition is that H is not of the form $H(a_0)$ where a_0 was engineered to make $H(a_0)$ yield E. What matters is therefore not the whole of B but only that part which was *actually used* in constructing $H(a_0)$; where the construction follows something like the logical pattern described above. The context of the discovery of a hypothesis H may therefore have a role to play in the 'justification' of H. It does not however suffice for a scientist to be *psychologically aware* of a factual result E in order for his conjectures H to become automatically adhoc with respect to E; he must furthermore have made some *objective use* of E in determining H. For example: Einstein surely knew about Michelson's results and about Mercury's perihelion before proposing his relativistic theories; but this in no way turns the latter into adhoc accounts of the known facts.

Finally: even if $H(a_0)$ were adjusted, in the manner described above, to yield E, $H(a_0)$ could still receive some measure of support from E; for we might, for logical reasons, altogether fail to find any a_0 such that $H(a_0)$ consistently entails E. This clearly happens whenever, for some K and for all Q, $K \cap Q$ is empty, i.e. $K \cap Q = \emptyset$. There exist two limiting cases in which $K \cap Q = \emptyset$ holds; namely those in which either K or Q is empty. Concerning $K = \emptyset$: consider the fictitious example in which E refers to a single instance of the addition law of velocities and H(a) to Special Relativity, where the speed of light a is provisionally treated as a finite adjustable parameter. Then no matter how we vary a, we shall always have $\vdash H(a) \Rightarrow \neg E$. I.e. $\vdash (H(a) \wedge E) \Rightarrow (a \in \emptyset)$. We can therefore take $K = \emptyset$. Note that we can infer $(\exists x)H(x) \Rightarrow \neg E$ from $H(a) \Rightarrow \neg E$; that is: $(\exists x)H(x)$ already constitutes a scientific theory since it possesses the potential falsifier E. As for $Q = \emptyset$, it represents the case where H(a) is compatible with, but also irrelevant to E; so no choice of a enables H(a) to decide E in a consistent manner. For example: take H(a) to be an economic theory in which we vary some parameter a; and let E be the same statement as in the last example. Since economic hypotheses do not bear on any physical laws, there will be no value of a such that $\vdash H(a) \Rightarrow E$. Thus $Q = \emptyset$.

The above method of selecting $a \in K \cap Q$ can therefore break down. It follows that H(a) can legitimately derive some support from the mere existence of just one element of $K \cap Q$. Remembering that $a = (a_1, \ldots, a_n)$ and $E \equiv e_1 \wedge \ldots \wedge e_m$, we conclude that the next—naturally unused— experimental result e_{m+1} might serve to test H(a) severely; for by hypothesis, H(a) will have been determined independently of e_{m+1}; so that if $\vdash H(a) \Rightarrow e_{m+1}$, then H(a) will have explained e_{m+1} in a non-adhoc way.

Generally speaking, strong evidential support goes hand in hand with overdetermination: if any m experimental results suffice to fix the n parameters $a_1, \ldots, a_n$ occurring in $H(a_1, \ldots, a_n)$, then any number of facts in excess of m can either refute or else strongly confirm $H(a_1, \ldots, a_n)$; for in this hypothetical case, any m data would uniquely determine $a_1, \ldots, a_n$; while the remaining facts could turn out to be either incompatible with, or else subsumed by $H(a_1, \ldots, a_n)$. In the latter case, $H(a_1, \ldots, a_n)$ will have been genuinely corroborated.

In conclusion, we can say that Poincaré's and Popper's value-judgments, though broadly correct, are too coarse-grained to capture the scientists' intuitive notion of corroboration; for even though H might have been doctored to yield E, such doctoring could in principle have proved impossible; which implies that since H has taken a risk—no matter how minimal—it is entitled to receive some measure of support from E.

1.7 The Problem of the Empirical Basis: Popper versus Poincaré

Against LeRoy's nominalism, Poincaré maintains that the scientist does not arbitrarily fabricate scientific facts; that the latter should moreover be regarded as translations into a more convenient language—more precisely: into the language of science—of socalled crude facts. Crude facts are expressed by propositions which, following John Watkins, we shall call autopsycholgical or level-0 reports. It is my opinion that these reports form the bedrock of all scientific systems. They are the simplest propositions which stand in need of explanation while not being capable, if taken by themselves, of explaining any other state-of-affairs (Watkins 1984, p. 79).

Let us provisionally refrain from any deep discussion of the meanings and referents of level-0 report. We shall simply remark that we are here confronted with yet another ambiguity in Poincaré's position; for where does the crude fact belong? Is it strictly phenomenological or does it also have an objective aspect? Is it purely subjective or does it refer to some mind-independent reality? In certain passages of his 1902 and of his 1906a, Poincaré claims that propositions like 'the current is on', which clearly have objective import, describe crude facts. In other parts of the same works, he maintains that a crude fact is unique since it comprises subjective perceptions which no proposition can individuate. This is so because our language has at its command "only a finite number of terms to express the shades, in number infinite, that my impressions might cover" (Poincaré 1906a, p. 117). Crude facts can nonetheless verify or falsify the level-0 propositions which are intended to capture some of their aspects. Given the individual character of each of our sensations, it seems to me that an objective basic statement like 'the current is on' cannot by right, that is by means of logic alone, be the mere translation into a convenient language of the observation report 'I see what seems to me to be the spot of a galvanometer move'. Only the crude visual fact, as distinct from the alleged passage of the electric current, can be indubitably ascertained. Yet an inductivist like Poincaré needs the *objective* basic statements in order to arrive at such experimental laws as 'All metals are good conductors'; for only the latter enable scientists to predict and hence to act.

We shall rationally reconstruct Poincaré's position by modifying it into what will henceforth be called the *phenomenological conception* of the empirical basis (or, for short,the *phenomenological thesis* or *phenomenological view*). This is defined as follows. We shall say of a singular sentence p' that it expresses a level-0 proposition if p' describes, in the first person, the immediate contents of the speaker's consciousness. Examples of such sentences are: 'I feel pain (now)', 'I seem to be perceiving a red patch', and so forth. It will be taken for granted that the truth-values of such propositions are *logically* independent of all transcendent states-of-affairs, that is, of all events occurring outside the speaker's consciousness. This is why level-0 statements are also referred to as immanent, autopsychological or phenomenologically reduced propositions.

According to the phenomenological thesis, all basic statements can ultimately be reduced to level-0 sentences; where 'reduced' does not mean 'made logically equivalent to'. More precisely: let p be a sentence such as 'the current is on' which describes a physical process; and let p' be the proposition, expressed in some phenomenological vocabulary, that a certain spot is seen to move. We shall regard p' as describing a sequence of perceptions correlated with p. It can be said that p' is the result of a phenomenological reduction which eliminates from p all references to any mind-independent reality. According to the phenomenological view, all experimental results can, in principle, be expressed by sentences having the same form as p'; they can thus effectively verify or falsify such sentences. The phenomenological thesis moreover presupposes the existence of a psycho-physical law A, such that: $\vdash A \Rightarrow (p \Leftrightarrow p')$. A will contain clauses both about the reliability of the instruments used during some experiment and about the observer's mental and physical state (for instance the precondition that he is not colour-blind). It is important to note that A should be regarded as undergoing empirical tests alongside every core hypothesis H to which A might be appended. A is therefore contingent, so that p is, not logically but materially equivalent to p'. As long as we accept A, we can assert $p \Leftrightarrow p'$ and hence also $p' \Rightarrow p$. If the experiment, having been carried out n times, yields the phenomenological results $p_1', \ldots, p_n'$, then in view of $p_i' \Rightarrow p_i$, and provided we can rely on all the level-0 reports, we are entitled to conclude that $p_1, \ldots, p_{n-1}$ and p_n are true. These last statements can then be generalised to yield some experimental law, P say. Induction is thus fallible on two counts. First: should A be false, then we can no longer infer p_i from p_i'; and so $p_1, \ldots, p_n$ need no longer hold good since one of our premises may be false. Secondly: even if A were true, P might still break down because it goes 'horizontally' beyond $p_1, \ldots, p_n$.

In order to test a complex system S one of whose components is A, one extracts a prediction q from the theory S, where the latter is taken in conjunction with a description p of the relevant boundary conditions. Thus: $\vdash S \Rightarrow A$ and $\vdash S \Rightarrow (p \Rightarrow q)$. (In fact, without the help of A, the core theory within S will generally entail $p \Rightarrow q$.) To p and q correspond the autopsychological sentences p' and q' respectively; so that: $\vdash A \Rightarrow (p \Leftrightarrow p')$ and $\vdash A \Rightarrow (q \Leftrightarrow q')$. It follows that: $\vdash S \Rightarrow (p' \Rightarrow q')$. (Note that in order to obtain this last implication, we need only assume $\vdash A \Rightarrow (p' \Rightarrow p)$ and $\vdash A \Rightarrow (q \Rightarrow q')$.) S is therefore refutable by means of propositions like $p' \wedge \neg q'$ whose truth-value can be effectively decided.

Let us compare Poincaré's phenomenological position—as thus reinterpreted—with Popper's views. Popper is prepared to accept $p \wedge \neg q$ as a potential falsifier whereas, according to the phenomenological thesis, only statements of the form $p' \wedge \neg q'$ are admissible as observation reports. But also contrary to Popper's standpoint, the phenomenological thesis implies that there is practically no possibility of error at the autopsychological level. (It could, more modestly, be shown that mistakes are *less likely* to occur at the observational than at the theoretical level).

Should it turn out to be correct, the phenomenological thesis would reinforce falsificationism at the cost of complicating the Duhem-Quine problem. Falsifications would no longer be a matter of convention; for we can now *ascertain* the truth of propositions of the form $p' \wedge \neg q'$ and hence the falsity of some global system under test. This would block the anarchist conclusions which Feyerabend justifiably drew from Popper's position (Feyerabend 1975, pp. 65–68). As a counterpart, the phenomenological thesis has to allow into every scientific system some psycho-physical laws linking physical processes to mental phenomena; which greatly increases the complexity of the Duhem-Quine problem.

Turning to Popper's demarcation criterion, one might be led to believe that the impossibility of verifying universal propositions flows from the physical impossibility of performing infinitely many tasks, while the possibility of falsification rests on the feasibility of finitely many observations carried out in order to decide the truth-value of a potential falsifier. Popper however goes out of his way to assert that observation—qua process grounded in perception—bears no epistemological relation to basic statements. Sense-experience may *motivate* our decision to accept a falsifier but it provides *no reason* for doing so:

> I admit again that the decision to accept a basic statement and to be satisfied with it is causally connected with our experiences—especially with our *perceptual*

> *experiences*. But we do not attempt to *justify* basic statements by these experiences. Experiences *can motivate a decision* and hence an acceptance or rejection of a statement, but a basic statement cannot be *justified* by them—no more than by thumping the table. (Popper 1959, p. 105)

According to Popper, basic statements can and generally should be objective propositions about the world. Not only are they theory-laden, they consist of low-level hypotheses containing dispositional terms; their verification is therefore impossible and it certainly has nothing to do with any observer's perceptions. A basic statement is acceptable if there is *intersubjective agreement* about its truth-value. This sort of acceptance can be revised and is therefore non-dogmatic; but when carefully examined, such 'revisions' simply mean that the previously accepted propositions, when taken in conjunction with other hypotheses, might be rejected as a result of fresh agreements about other 'falsifiers' which are also *conventionally* accepted. At no point does this potentially infinite process involve any genuinely *epistemological* considerations, in other words any considerations linking the truth-value of the basic statement either to the act of observation or to that of reaching a consensus.

This view of the empirical basis threatens to destroy the presumed asymmetry between verification and falsification; for we now have a proposition, dubbed 'theory H', confronting another proposition which we call 'basic statement B'. Despite these labels, all we are entitled to assert is the logical incompatibility of H and B: all we know is that $\vdash \neg(H \wedge B)$ holds. The relationship between H and B is therefore perfectly symmetric. Popper admittedly describes B as a *low-level hypothesis,* but this is of no great help; for since Not $\vdash (H \Rightarrow B)$ and Not $\vdash (B \Rightarrow H)$, the levels of H and B are not comparable except in some vague intuitive sense. More seriously: nothing in Popper's analysis tells us that the hypotheses impregnating B are less risky than H. All we can truly claim is that over the past four hundred years or so, science has pursued a largely empiricist policy, that scientists have reached agreement over propositions resembling 'B' rather than over statements having superficially the same form as 'H'. But as conceded by Popper himself, we can offer no scientific explanation of the social success of this empiricist strategy (Popper 1979, p. 132). Worse still, we cannot explain why this policy issued in great technological breakthroughs. The latter are taken to depend on the truth, the probable truth, or the approximate truth of some consequences of our theories, or at least on our having, by and large, rejected false hypotheses rather than accepted false basic statements. Yet truth-considerations have so far played no role in Popper's methodology, so that any sustained technological progress becomes an ongoing miracle. Needless to say, Poincaré would have rejected this 'miraculous' interpretation out of hand.

Popper's view of basic statements, which will be called the *conventionalist thesis* or *conventionalist view,* was in fact shared by some members of the Vienna Circle like Otto Neurath. This thesis is clearly sociological; and it is so because a transcendental critique has allegedly shown its psychologistic rivals to be incompatible with successful scientific practice.

It has to be admitted that there are seemingly solid reasons for rejecting level-0 sentences as suitable candidates for membership of the empirical basis. Popper argued as follows. Level-0 statements are incorrigible only in the sense of being *psychologically* indubitable; but we know that feelings of conviction, no matter how strong, are often misleading: they neither establish nor even probabilify the propositions at which they are directed. Autopsychological reports suffer from the added disadvantage of being private to the person uttering them. Not being susceptible of any repeated testing, their veracity cannot be independently checked (Popper 1959, p. 93). Hence no intersubjective agreement regarding their truth-value or their general acceptability can be consistently reached. Popper thus holds a *psychologistic view* of level-0 reports which are taken to rest exclusively on the feelings of conviction of those who, while uttering their autopsychological statements, are *psychologically* incapable of doubting them.

Watkins put forward another criticism of the phenomenological thesis, the so-called *unmanageable largism objection:* no matter how much we strengthen a scientific system, say by adjoining various auxiliary hypotheses to its hard core, it will never be powerful enough to entail any immanent propositions; which is why we should settle for an empirical basis consisting of objective and hence fallible level-1 statements (Watkins 1984, p. 247).

If Poincaré's phenomenological position—as rationally reconstructed—is to be rescued, then both Popper's psychologistic view and Watkins's unmanageable largism objection must be rebutted.

1.8 Refutation of the Psychologist Thesis

There are admittedly sound *historical* reasons for holding a psychologistic view of autopsychological reports. This last phrase anyway has a pleonastic ring to it. As explained in Chapter 5 of Popper's *Die beiden Grundprobleme der Erkenntnistheorie*, Fries realised that on pain of falling into vicious circularity or into some infinite regress, we have to give up the idea of justifying *all* synthetic propositions. In order to avoid dogmatism, Fries accepted experiential statements as something which, as a matter of brute psychological fact, we cannot help believing in. Thus Fries openly subscribed to the psychologist thesis.

Franz Brentano, a contemporary of Poincaré as well as a compatriot and near-contemporary of Popper, tried to account for the certainty of level-0 reports in non-psychologistic terms. He thereby developed a so-called 'Evidenzlehre' which was meant to provide an objectivist account of self-evidence (Brentano 1930, Section 4). Brentano distinguished between the blind compulsion which misleads us, even in cases of hallucination, into believing in the existence of allegedly external but unlikely objects; and genuine self-evidence, which is the objective hallmark of indubitable propositions like the logico-mathematical principles, the axioms of probability, and certain autopsychological reports. The latter were called statements of *secondary* or *inner* perception, but since they were known to be contingent, they could not be established without recourse to some synthetic principle. Once again, the twin problems of infinite regress and of vicious circularity reared their ugly heads. Brentano envisaged two solutions to these problems. The first consisted in giving a laundry list of propositions which were simply declared self-evident. This came down to postulating what was needed and, as is well known, represented all the advantages of theft over honest toil. According to the second 'solution', self-evidence can be infallibly recognised through direct insight. In other words, though self-evidence be an objective attribute of propositions, we have to rely on psychologistic criteria in order to identify it. Thus Popper can be excused for taking such 'Evidenzlehren' to be thinly disguised forms of psychologism. I think nonetheless that Brentano's thesis can be rescued by methods somewhat similar to those used by one of his followers, Anton Marty. We have already mentioned that in trying to establish the truth of level-0 propositions, we have to stop somewhere; moreover, that this 'somewhere' must be synthetic and hence questionable. So why not stop, as Brentano suggested, at the level-0 reports themselves? Such reports could simply be dubbed self-evident. But the point at which the analysis should be broken off is not a matter of indifference: as regards the certainty of autopsychological statements, we might in the end succeed in determining an underlying principle which links up with other components of Poincaré's or of Popper's philosophical system. In other words, although some unquestioned principle must finally be invoked, the latter may be needed in other areas of a global position which thereby becomes more unified.

In what follows, I propose to show that the certainty of level-0 sentences rests on an intuitive version of the Correspondence Theory of Truth; and it is significant that Brentano resorted to his 'Evidenzlehre' precisely when he started having doubts about the Correspondence Theory. Note that Truth-qua-Correspondence became a central theme in Popper's later philosophy and that at least with respect to 'crude' factual statements, Poincaré subscribed to an intuitive version of the Correspondence Theory. Thus my move will, I hope, demonstrate that the phenomenological view forms a natural complement both to Popper's and to Poincaré's methodologies.

I shall try to show that level-0 reports can be recognised as true, not by being derived from self-evident propositions, but through our having a direct access both to their referents and to the meanings of the predicates occurring in them. Only in such epistemologically privileged situations can we ascertain a truth-correspondence between a sentence and the state-of-affairs it supposedly signifies. We must therefore start by analysing what kind of referents autopsychological propositions can be legitimately claimed to possess. Ayer wrote:

> The next step, continuing with our example, is to convert the sentence 'it now seems to me that I see a cigarette case' into 'I am now seeing a seeming-cigarette-case'. And this seeming-cigarette-case, which lives only in my present experience, is an example of a sense-datum . . . the expression must be used in such a way that the experience of the physical object which appears to be referred to remains an open question: there is no implication either that it does exist or that it does not . . . What appears most dubious of all is the final step by which we are to pass from 'it seems to me that I perceive x' to 'I perceive a seeming-x', with the implication that there is a seeming-x which I perceive . . . For to talk of someone's sensing a sense-datum is intended to be another way of saying that he is sensitively affected; the manner in which he is affected reappears as a property of the sense-datum. . . . (Ayer 1956, pp. 106, 111, 115, 118)

In these passages, Ayer mentions one of the central problems with which Brentano struggled throughout his life. Brentano laid stress on the intentional structure of consciousness, which is always a consciousness *of something*. Every mental act intends, that is: it is directed towards some object or other. Since these objects need not necessarily exist independently of the mental operations generating them, Brentano called them 'intentional' or 'immanent' entities. This does not entail that no transcendent entity exists over and above the intentional one; but only that even in the absence of the former, the latter is always present. Given the phenomenological continuity between hallucination and veridical perception, all perceptual objects belong to the same category and must therefore be non-physical. The above quotation shows that Brentano's intended perceptual objects can be identified with Ayer's sense-data. But can immanent entities be said to exist in any legitimate sense? Note that the answer must be in the affirmative if 'I see a seeming-x' is to count as a true level-0 report.

Initially, Brentano half-heartedly ascribed to intentional entities a kind of existence which he misleadingly called 'inexistence in consciousness'. Sense-data were not supposed to exist in the ordinary realist sense; but despite being private to one mind, they were taken to be *things* which functioned as vehicles for such phenomenal properties as colour, smell and shape. They were interposed between the mental acts on the one hand, and the external world on the other. They could—in principle—reflect the properties of transcendent objects but they existed only 'in the mind'. Should this thesis prove tenable, then we could identify level-0 statements with propositions of the form 'I perceive a seeming-x' where x is some description. The referent of 'seeming-x' would thus be a sense-datum. Since the latter is an immanent entity, we can directly apprehend it and check whether it has the phenomenal properties expressed by x. Level-0 statements would thus have ascertainable truth-values.

With time, Brentano grew dissatisfied with his shadowy intentional realm; and this for a variety of reasons. To begin with, he became a wholehearted empiricist and hence doubted the existence of such ethereal entities as sense-data. Later, he also felt that the postulation of objects of thought would inevitably engender logical paradoxes. He was afraid of being driven to accept the existence of logically impossible objects like round squares. In this, he was undoubtedly mistaken: there is of course nothing to prevent us from conjoining squareness with roundness and from thereby obtaining the empty class ø; but since we can neither perceive nor even imagine round squares, the principle of contradiction is not threatened by the introduction of an abstract object of thought like ø. All the same, Brentano seems to have had an unerring instinct regarding the logical inadmissibility of intentional entities; for these are both richer and, in other respects, less determinate than ordinary objects. They are richer in that they present themselves as theory-laden, with the result that the same physical object can give rise to infinitely many different sense-data. Sense-data have a focus fixed by the observer's attention in such a way that the same chest X-ray can, for example, be integrated in infinitely many ways. Moreover, two-dimensional objects like a painting or an image on the television screen are instinctively perceived as possessing depth. Only through considerable effort can I bring myself to 'see' them as two-dimensional pictures. By contrast, the photo of an object taken from an unusual angle will initially appear as a two-dimensional arrangement of colours and shapes; the 'real' object may then be identified and the new sense-datum will not only supplant but also completely suppress the old one. In the case of a dot on a radar screen, both the layman and the expert will be aware of a two-dimensional picture, to which only the expert will add some information about the distance of a flying object; and only through an act of will can he suppress this theoretical admixture and 'retrieve' a purely visual image. Every transcendent object may thus give rise to an indefinite number of sense-data. Accepting the actual existence of the latter will therefore lead to ontological inflation. But there are worse dangers still; for intentional entities are also less determinate than their physical counterparts: they possess only those properties which are explicitly projected onto them by our mental acts. A sense-datum is an exclusively phenomenal entity; it is some *object as perceived or as thought.* A speckled hen is apprehended as having a finite number of speckles but neither as having 100 speckles nor as not having 100 speckles. Should we transform this innocuous sentence into a statement about the sense-data themselves, then we would clearly violate the law of the excluded middle: the phenomenal hen will have a finite number of speckles; yet it is not the case that this number will be either equal to or different from 100. In this respect, the hen is on a par with Hamlet who cannot be said to be, or else not to be, 1.75 metres tall; for Hamlet was simply not intended to have any specifiable height. An intentional entity will therefore lack every property—together with its negation—which is not explicitly denied or attributed to it by some act of consciousness: Husserl expressed this principle by asserting the existence of a strict correlation between the noesis on the one hand, and the noema, taken together with its intentional object, on the other. I take this to mean that all talk of intentional entities can be considered as shorthand for certain descriptions of the underlying mental operations. One can of course decree that even sense-data must obey the law of the excluded middle; but they will then escape the mind's control and cease to be immanent entities. We may, for example, be unable to count the number of speckles on the 'phenomenal hen'; and any mechanical procedure for counting them will alter the corresponding sense-datum. Should the hen moreover have been hallucinated, then to assume that it ought still to carry a well-defined number of speckles is nothing short of fantastic. Thus, for any intentional object y and any property P which can be meaningfully predicated of y: P(y) will hold iff y is intended—by the corresponding noesis—to have P; y may thus fail to possess both P and ¬P.

Brentano consequently searched for a different solution which he found at the turn of the century, and then kept unaltered until his death in 1916. Before discussing this solution in any detail, let us examine a concrete example. Consider the (truthful) proposition 'I see a red house' and suppose that there is no physical house out there at all; in other words, assume that I am hallucinating. If 'red house' be given its ordinary realist meaning, then my statement will clearly be false. It could conceivably be turned into a true sentence if 'red house' were to denote a sense-datum. But as already explained, we have good reasons for denying the existence of all intentional entities; so the truth-value of the proposition becomes problematic. Fortunately, the sentence 'I seem to be perceiving a red house' has a truth-value which is independent of the existence or non-existence of the referent of 'red house', and this is all we need in order to secure a phenomenological basis for the sciences.

According to Brentano's final solution, all forms of consciousness are described by statements of the form 'I seem to be perceiving a red house'. In other words, consciousness always possesses two indissolubly connected components; it is "zweistrahlig"; that is: it consists of two rays. There is first a primary conscious act directed at a primary object, such as the red house. In no sense however can a primary intentional object be said to exist qua sense-datum; there simply is no immanent red house. Only the ego with its various directional activities exists; and all statements about the phenomenal red house are shorthand for descriptions of modes of the underlying conscious activity. Secondly, the primary act is always accompanied by a so-called inner or secondary consciousness which intends the primary activity as its immanent object; the latter exists in the ordinary sense and through it, secondary consciousness is 'obliquely' directed towards the non-existent primary object, for example towards the red house qua sense-datum. Thus, whenever I primarily seem to be perceiving a house which is red, I am at the same time secondarily aware of so seeing it. This inner awareness is finally aware of itself as well as of all primary conscious acts. According to Brentano's definitive view, every conscious act is therefore such that:

(1) There exists a primary conscious activity directed at a non-existent primary intentional object (which does not of course imply that there is no corresponding transcendent object).

(2) There is a secondary conscious activity which directly targets the (existing) primary act and through the latter, *obliquely* or *indirectly* points to the (non-existent) primary object.

(3) Secondary consciousness is finally aware of itself. Thus tertiary and all higher levels of consciousness collapse back onto the secondary level.

There is clearly something wrong with this Brentanian solution, but I think that it can be altered into an acceptable thesis. Note first that level-0 reports should be put in the form 'I seem to be perceiving x' rather than 'I perceive a seeming-x'. We are thus dealing with *descriptions* of primary acts which are issued by secondary consciousness. We can therefore abstain from discussing the ontology of sense-data, or in other words, of seeming-x's. Secondly, however, we find Brentano defending the view that *every* mental act must be accompanied by secondary consciousness and should therefore be self-aware. I find this thesis unconvincing. Not only would it exclude all subconscious activity; it also entails that animals must either be capable of self-consciousness or else possess no consciousness at all. There is also some direct evidence—supplied by memory—that in everyday life we are not always reflexively conscious of perceiving something; we do not always watch ourselves observing but straightforwardly perceive objects which we take to be external to our minds; that is, our attention is directed exclusively at *primary* states-of-affairs. It is however also true that through reflection, the primary acts themselves can be made into the existent objects of higher mental activities; that is, of inner or of secondary consciousness. This gives rise to level-0 reports like 'I seem to be seeing a red house' and 'I believe there is a red house in front of me'; in such cases, the immanent intentional objects, namely my seeing what I take to be the red house and my belief in its physical existence, come under the scrutiny of secondary consciousness and are thus known to 'exist' in the ordinary sense of this word. Having admitted that a conscious act can occur at a definite level without necessarily involving any higher one, we no longer need Brentano's assumption (3) which was clearly introduced in order to block an infinite regress: if every conscious level presupposed a higher one, then we should have to accept an actual infinity of strata in every mental act; which is patently absurd. Through identifying the secondary with the tertiary level, Brentano arrested this regress at the secondary stage. We can however admit that every level could *but need not* be made into the existent object of some higher level activity. Everyday experience occurs exclusively at the primary level and gives rise to statements like 'Here is a glass of water', which is fallible since 'glass' and 'water' are given their usual transcendent meanings. All levels above the primary one have immanent and hence necessarily existing intentional objects. More precisely, once the higher act is performed, then its object must exist and the corresponding *descriptive* proposition becomes incorrigible. Of this kind are the level-0 reports which describe the findings of secondary consciousness. Thus the testing of scientific theories takes place *at the level of inner consciousness* which issues in statements of the form: 'I seem to be observing a glass of water (that is, a glass containing what I take to be a liquid composed of H_2 and of O)', or 'I seem to be seeing a red house'. Note that primary consciousness remains dependent on the theories entertained by the observer; but the role of secondary consciousness is exclusively *descriptive* of these primary modes. As for philosophical reflection, it is carried out at the tertiary level or even above it; for it examines, among other things, the relationship between the primary and the inner activities.

Before accounting for the infallibility of level-0 reports, let me dispel the myth that propositions can be justified only by means of other propositions. Once a statement q has been (validly) deduced from another, p, say, then the truth of p will naturally entail that of q; for by definition, all deductions transmit truth from premises to conclusions. The myth just referred to asserts the converse of this trivial deductive principle: it maintains that all we can know with any certainty is the transmission of truth, not the truth of any nontautological propositions. Of course: if one subscribes to the Correspondence Theory, then no truth-correspondence can ever be infallibly ascertained between a realist hypothesis and the *transcendent* states-of-affairs it is meant to denote; for the external world is something we do not directly apprehend; we conjecturally aim at knowing it, not by acquaintance, but through the medium of theories which cannot be effectively verified. This conclusion does not however apply to the relationship between level-0 statements and the *immanent* states-of-affairs they describe. There are in fact two features of autopsychological propositions which make them epistemologically privileged:

(a) They signify states-of-affairs which German-speaking philosophers appropriately call *Erlebnisse.* As their name indicates, *Erlebnisse* can be 'lived through', so that we have an unmediated access to them. We can thus check any level-0 statement against its referent. I can, for example, directly compare the sentence 'I seem (or seemed) to be seeing a red house' either with a present or with a remembered experience. But I am in no position to apprehend directly the state-of-affairs described, say, by the statement 'There exists an electromagnetic field in this room'; for I am in principle unable to leap out of myself and partake of the state of the field. My privileged access to the referent of an autopsychological statement does not yet guarantee the latter's infallibility; for as already indicated, the referent may be a remembered and hence a misremembered one. Whence the necessity of a second condition.

(b) Through inner or secondary consciousness, we can become aware of our mental acts *as we perform them.* I can observe, or more accurately: I can sense myself observing some primary object. This either eliminates or greatly reduces the possibility of my being mistaken in claiming, for example, that I *now* seem to be perceiving a red patch; but only *with the proviso* that the claim be made simultaneously with the experience it describes. Without having to rely on my fallible memory, I can thus directly compare a report about my *present* state of consciousness with its presumed present referent. This simultaneity also ensures that all the descriptive terms occurring in my statement are given the meanings intended at the time of my uttering it. A truth-correspondence can thus be immediately ascertained.

As remarked by Ayer, verbal mistakes are always possible. We often shout 'Right' when we ostensibly mean 'Left'. I may say 'I seem to be seeing a purple patch', where by 'purple' I now mean what I previously meant by 'red'. These are rather trivial instances of error, but there are more puzzling ones. Suppose that I now see a red patch; suppose further that, yesterday, I saw a coloured patch which I correctly described as purple but which I now wrongly remember as having had a colour identical with my present colour-sensation. I nonetheless rightly remember using the word 'purple' yesterday. As a result, I now claim to be perceiving (or rather to appear to be perceiving) a purple patch. It therefore looks as though I were mistaken in a non-verbal way. An illusion of error however arises only because no proper phenomenological reduction has been carried out. In 'I now seem to be perceiving a purple patch', 'purple' has been surreptitiously given a transcendent meaning not intended by the speaker. His sentence, when properly unpacked, reads: 'I (now) seem to be perceiving a colour which (now) seems to be identical with the colour which I (now) recall having seen yesterday'. This report is properly of level-0 in both senses (a) and (b) and it is certainly true; it being understood that 'recall' is not to be regarded as a success word.

The two points (a) and (b) have been conflated by many thinkers, though not by Brentano. He even began by holding that we are in principle incapable of observing ourselves observing, that we consequently have to rely on our memory in order to reach any convincing phenomenological conclusions (Brentano 1924, Chapter 3). Although Brentano later rightly revised his views, his first thesis points to a valid *logical* distinction between (a) and (b) (Brentano 1968, Chapter 1). According to (a), if we want to determine the truth-value of a statement, not by inferring it from other propositions but through confronting it with its presumed referent, then we ought to restrict ourselves to autopsychological reports irrespective of whether these describe our past or our present experiences; for we can directly apprehend only our own mental states. As for point (b), it warns us that should we want to eliminate—or greatly diminish—the risk of error, then we should furthermore limit ourselves to descriptions of our *present* mental states; for memory is known to be highly fallible.

Let us now apply these conclusions to the falsificationist thesis. In order to test a scientific law, it is no use appealing to any of its consequences which describe mind-independent events; for our access to these is exclusively theoretical. In other words: we cannot break out of the circle of theories—and of yet more theories—by remaining within the transcendent domain. We have a double entry only to those propositions which flow from a scientific system and can furthermore be directly checked against the states-of-affairs they purportedly signify. Such statements can only be the level-0 reports discussed above. Even if we feel unhappy or unsure about their incorrigibility, point (a) demonstrates that they are anyway the only candidates for the status of basic statement. I have moreover tried to show that their certainty is not founded on any observer's feelings of conviction, but on the possibility of checking them against their referents (from such a successful check, a feeling of conviction may naturally arise). The psychologist thesis about autopsychological statements is therefore false.

1.9 Unmanageable Largism?

There remains one major difficulty confronting Poincaré's phenomenological thesis: can autopsychological predictions be *logically* derived from any—appropriately extended—scientific systems? John Watkins has denied this possibility by arguing that a level-0 empirical basis would lead to unmanageable largism. To my mind, Watkins's criticism constitutes the most cogent objection ever brought against the phenomenological thesis (Watkins 1984, Chapter 7). I shall nonetheless try to counter it.

It has already been conceded that the derivation of autopsychological reports greatly increases the complexity of the Duhem-Quine problem. For in deducing level-0 statements from scientific premises, not only do we have to consider a central theory taken together with auxiliary hypotheses and the relevant boundary conditions; we also have to add premises about the effect which the belief in certain laws has on the experimenter's state of mind and hence on the conclusions he will draw from his observations. To the boundary conditions we therefore have to adjoin assumptions about the observer's mental and physical states. For example, if our test involves the recognition of certain colours, we may have to presuppose that our experimenter is not colour-blind. So why bother with these complications which led Duhem, Popper, and Watkins—but neither Poincaré nor Einstein—to propose keeping autopsychological reports out of the domain of physical science? The answer has already been given above and is really quite simple: in case of conflict with experience, we want to have some good reason—rather than a mere motive—for holding our premises to be false rather than simply inconsistent with a transcendent statement which we collectively—but also provisionally and highly fallibly—choose to accept. As for the non-feasibility thesis, let me first say that the onus of proof rests on the opponents of the phenomenological view. Using Popper's transcendental method of criticism, I shall show that the actual practice of science presupposes, if only implicitly, the phenomenological thesis. Two simple examples will make the point. We all agree that a colour-blind experimenter's report can be discounted in test-situations where the perception of colour matters. Yet a person's inability to distinguish between various colours is not relevant, say, to the objective blueness of a piece of litmus paper but to the recognition of such blueness. Whether we are aware of it or not, the basic statements we actually rely on have therefore the same form as 'This, which I take to be a piece of litmus paper, looks blue' rather than 'This piece of litmus paper is blue (that is, it reflects light of a certain frequency)'. This last objective proposition can in fact be deduced from hypotheses which make no reference at all to any conscious observer. It follows that such a derivation will not be *blocked* by altering assumptions about the absence of colour-blindness, for no such assumption figures among the premises. Only the derivation of the corresponding autopsychological report requires the premise that the observer should not be not colour-blind; which is precisely why ascertaining colour-blindness could save our core theory from refutation. Thus only the phenomenological thesis renders the practice of science rational. Similar conclusions apply to the X-ray example mentioned above: the actual existence of an abnormality on the X-ray in no way depends on the observer's state of knowledge; yet we feel justified in ignoring any experimental account which is given, not by a trained doctor, but by a layman. Some inferential link must therefore exist between a description of the observer's state of knowledge and his report. It follows that the latter cannot possibly be an objective statement about some transcendent state-of-affairs; it must rather assume the form 'I notice such and such a patch which, for the following reasons, I take to be evidence for the presence in the lung of active Koch bacilli . . .' This statement cannot be inferred without recourse to some hypothesis about the observer's professional background; which is why a layman's report constitutes no test of any system comprising a theory about some illness and certain assumptions about the experimenter; for the layman fails to satisfy these extra assumptions.

Repeatability and intersubjective agreement, or *consensus,* ought not therefore to be built into the definition of a basic statement. From a *phenomenological* viewpoint, we can nonetheless see why in presumed cases of refutation, the actual repetition—by different people and under differing conditions—of some falsifying experiment makes perfect sense. The main purpose of repeating a test is to render highly improbable a series of coincidences which might have falsified some assumptions about the non-existence of certain factors or about the experimenter's mental and physical health. In this way, repetitions enable us to hive off a number of peripheral premises and hence to blame a refutation on some core theory. The Duhem-Quine problem is thus made more manageable and the refutation of isolated hypotheses possible, albeit in a fallible-conjectural way. Popper was therefore right to insist on the requirements of repeatability and intersubjective agreement, not because these conditions are defining characteristics of basic statements, but because they constitute a good method of dealing more effectively with the Duhem-Quine problem.

Let me end this chapter by putting forward a simple logical argument in support of the phenomenological approach which was advocated—if only implicitly—by Poincaré. As long as level-0 propositions are regarded as explananda and as long as we subscribe to hypothetico-deductivism, the only way such explananda can be explained is through their being deduced from a set of assumptions. By the completeness theorem, the latter can be taken to be finite. Each of these premises can then be either derived from or else adjoined to some core hypothesis, where the latter is naturally taken in conjunction with descriptions of the relevant boundary conditions. In this way, we are bound to obtain a finite system which yields the level-0 statements as deductive consequences. Whether or not one chooses to call such a system 'scientific' is a terminological matter. The system will certainly be testable against autopsychological reports, and it will consequently be refutable in the strong sense. This is why it seems to me that Watkins, who so forcefully underlined the epistemological relevance of level-0 statements, ought also to have followed Poincaré in accepting these statements as constituting the empirical basis of *all* the sciences.

2

Poincaré's Structural Realism and His Logic of Discovery

So far we have addressed logical and methodological but no genuinely metaphysical questions. Twentieth-century empiricist philosophers regard metaphysics as having played—at best—a *motivational* role within the sciences: like art and poetry, metaphysics may inspire a scientist but it does not provide him with any meaningful information. In the present chapter, I propose to refute this empiricist thesis by demonstrating that realism—in its *structural* form—played a central *heuristic* role both in Poincaré's epistemology and in his logic of scientific discovery.

2.1 Realism and the Reality of Universals

Poincaré's structural realism is in a sense a reversal of Quine's slogan: to be is to be quantified over (in some first-order theory); for according to Poincaré, only the universals, and more particularly the relations occurring in a unified and empirically successful theory, mirror the ontological order of things. As for the nature of the relata, it will forever remain hidden from us.

Let us briefly examine Quine's philosophical position which closely follows Tarskian semantics, that is: the semantics of first-order languages where only the individual variables are allowed to occur bound (Quine 1980b). The domain, the basic ontological layer of any interpretation J of a first-order language, is some set D of objects. D is the class over which the individual variables range. Every name belonging to the primitive vocabulary denotes an element of D. It is well known that both names and function letters can be eliminated through the introduction of new predicate letters. Thus only the latter—which express relations—are indispensable; each one of them corresponds to some subset of the Cartesian power D^n of the basic domain D; where n refers to the number of argument-places of the predicate in question.

Quine holds that there exists a fundamental asymmetry between the individual names on the one hand, and the predicate constants on the other: whereas the former denote elements of the domain D, the latter do not, strictly speaking, refer to anything at all; in other words: an n-place predicate does not name the attribute or subset of D^n which ought to correspond to it under J. We are consequently committed to the existence of no such subsets, the reason being that we have no predicate *variables* P which would enable us to assert the existence of at least one P possessing some property *f*. That is, $(\exists P)f(P)$ is inexpressible in a first-order theory. According to Quine, the existence of an object of any type must therefore be warranted by the ordinary linguistic locution: 'there exists some P such that . . .', where P is a variable of the considered type. Quine concludes that the subclasses of D^n do not have the same claim to existence as do the elements of D. If we moreover accept the Quinean thesis that by virtue of its completeness, first-order logic is the only *logic* worthy of the name, then we must conclude that no universals but only individuals exist; or at least that the former are metaphysically more real than the latter.

Even if we keep to first-order logic, this Quinean theory of ontological privilege proves difficult to defend. It seems however less indefensible than its opposite, which is precisely Poincaré's structural realist thesis; namely that in a unified and highly confirmed theory, the basic relations mirror some objective reality while the relata remain inaccessible to human knowledge. Though prima facie hard to vindicate, this view appears to me to be on the right track. The difficulty in defending it stems from the fact that classical semantics seems unable to interpret relations except through their relata. More precisely: in ascribing a truth-value to 'R(a,b)', we have first to determine the referents of both 'a' and 'b' in order then to ascertain whether <a,b> belongs to the referent of 'R'. Yet according to structural realism, we often have a good reason for supposing that 'R' reflects a real connection between elements about whose intrinsic nature we know next to nothing. The conditions under which we are entitled to make such a realist claim obtain whenever we have a highly unified hypothesis H which both involves R and explains a whole host of seemingly diparate facts in a non-adhoc way. (Needless to say, the explanation of these facts must bring R into play.)

Given the absence of any structural semantics on a par with the classical one, any defence of structural realism will inevitably involve a sort of hand-waving together with a reliance on specific examples which 'exact' philosophers contemptuously dismiss. I shall not however allow myself to be deterred. The relation 'R' mentioned above will typically contain mathematical predicates whose form, thanks to the Correspondence Principle, remains approximately constant across successive revolutions. By contrast, the intended referents of R's arguments may dramatically change during periods of scientific upheaval. This is illustrated by the following examples, some of which were adduced by Poincaré himself (also see Worrall 1996). Though Fresnel's, Maxwell's, and Lorentz's hypotheses seemingly refer to very different entities, namely to a mechanical medium, to the electromagnetic ether, and to some disembodied field respectively, the basic equations are approximately the same in all three cases. Another example is afforded by Newtonian mechanics which was altered by the transition from classical to quantum physics: whereas Newton's laws, namely: $p_x = m \cdot dx/dt$, $dp_x/dt = -\partial V/\partial x$, and so forth, connect the acceleration of a single particle with the net force acting on it, the corresponding quatum-mechanical equations: $<p_x> = m \cdot d(<x>)/dt$, $d(<p_x>)/dt = -<\partial V/\partial x>$, etc. are relations between various expectation values, that is: between quantities involving probabilities. By virtue of the Correspondence Principle, the two systems of equations, classical and quantum-physical, nevertheless have the same form. In all these cases, the relata appear to have changed dramatically while the relations themselves were left almost intact by successive scientific revolutions. As for the confirmed empirical content of successive hypotheses, it has steadily increased. That it could not decrease is again warranted by the Correspondence Principle (see below, Chapter 4).

This quasi-cumulative aspect of progress, coupled with the dramatic predictive success of the sciences, led Poincaré to formulate his no-miracles argument: short of accepting an implausible conspiracy theory, the development of science must be regarded as indicating that relations common to successive theories reflect objective aspects of the ontological order, but we have no reason to suppose that the individual terms occurring in our theories denote separate items of reality.

Paradoxically, Poincaré's brand of conventionalism proves more realistic with respect to universals than with respect to individuals. And once again, the semantic difficulties mentioned above arise: how can we rationally maintain that some relation holds good when we totally ignore what its arguments denote? We might answer that these arguments actually *possess* separate referents, about which nothing is known beyond their satisfaction of the relation in question. The problem with this minimalist claim is that even the logical types of the relata are often uncertain. For example, in going from classical to quantum mechanics, we find that although the form of the Hamiltonian remains unchanged, its argument-places are no longer filled by quantities denoting positions, times, and momenta but by

Hermitean operators, that is: by second-order entities. We must therefore remain agnostic about the nature of the relata while making some epistemic claims—no matter how modest—about the relations themselves. And we have already noted that no coherent semantics for this approach is as yet available (except possibly in Category Theory). Fortunately for the structuralist, mathematics provides examples where the focus shifts from the study of elements subsumed under certain predicates to higher-order relations between the predicates themselves. For instance: sets, properties of individuals, bear to one another certain relations which are directly investigated by Boolean Algebra. In other words, Boolean Algebra concerns itself with classes without explicitly taking account of the elements of these classes. Similarly, topologies consist of systems of open sets, where the latter have properties and bear to one another relations which can be examined abstractly, that is: independently of the membership relation between point and set; and quantum theorists extract non-classical logics from the lattice of subspaces of a Hilbert space, *without* mentioning the vectors constituting these subspaces. Lattice theory can thus be viewed as a study of universals without direct reference to any individuals. (For an excellent study of Boolean algebras and, more generally, of lattice structure, see Hermes 1955). Such theories can in turn be expressed in first-order languages whose variables range over certain domains of objects; but these objects are universals rather than genuine particulars. Even from the semantic viewpoint, universals can thus be apprehended directly, that is: without specifically mentioning the indiviuals they subsume.

These considerations lead to another point highly relevant to Poincaré's philosophy. A physical theory is a nexus of first-order relations among individuals. Through shifting the focus away from the individuals and towards the relations and their second-order properties, Poincaré was led to devise a number of heuristic constraints to be imposed on the form of first-order relations. This is hardly surprising since all heuristic reasoning is *metatheoretical* or second-order in character. Poincaré's logic of discovery largely consists of the following principles:

(a) the Correspondence Principle, together with the requirements

(b) of Lorentz-covariance,

(c) of derivability from a principle of least action, and more particularly,

(d) of the posibility of a Hamiltonian formulation of the laws to be constructed.

Poincaré equated the existence of Hamiltonian functions with the possibility of a mechanical reduction of certain branches of physics. What is relevant here is not the correctness or incorrectness of this mechanical thesis, but the central role played by heuristic considerations in arriving at what Poincaré called the physics of principle. This consists of very general hypotheses whose mathematical form reflects high-level regularities in nature. A good example is provided by the conservation laws, which are known to be closely linked to Hamiltonian principles.

By using the heuristic methods sketched above, Poincaré determined the relativistic transformation rules for the electromagnetic field, thereby correcting a major error made by Lorentz in transforming the charge density. More importantly: Poincaré described the structure of the Lorentz-group within a four-dimensional continuum; he then used this group in order to construct a Lorentz-covariant gravitational theory which yields Newton's inverse-square law as a limiting case for small velocities (see Zahar 1989, Chapter 5; and see below, Chapter 4). It can moreover be legitimately claimed that the Hamiltonian approach advocated by Poincaré, as distinct from the strictly Newtonian methods of description, enabled scientists like Schrödinger and Dirac to effect a transition from classical to quantum mechanics; and that a combination of the heuristic principles proposed by Poincaré, namely: the requirements of Lorentz-covariance and of the existence of a Hamiltonian, led Schrödinger and then Dirac to their celebrated relativistic equations for the electron. After the wave-mechanical revolution of 1925–26, it was realized that the Hamiltonian had to be turned into an operator according to the following rules: $q \to q^*$; $t \to t^*$; $p \to -i\ (h/2\pi)\partial/\partial q$; $E \to (ih/2\pi)\partial/\partial t$; where: q and t are respectively the space and the time coordinates; p is the momentum conjugate to q; and q^*, t^* are the multiplicative operators defined by: $q^*(\psi) = q \cdot \psi$ and $t^*(\psi) = t \cdot \psi$ for all functions ψ. The transformation: $[H(q,p,t) = E] \to$

$[H(q^*, -i(h/2\pi)\partial/\partial q, t^*)\ \psi = i(h/2\pi)\partial\psi/\partial t]$ illustrates Poincaré's thesis that the form of the function H matters more than the nature or even the logical type of its arguments. Together with Einstein, Poincaré can therefore be regarded as the founding father of modern physics. (For a more formal approach to structural realism, see below, Appendix 4.)

2.2 Interpretation versus Reduction: Duhem's and Poincaré's Views about the Structure of Physical Theory

Poincaré's structural realism is prima facie similar to some of Duhem's theses; but unlike Poincaré, Duhem actually gave a coherent account of his philosophy of science (Duhem 1914). I hope that the following critique of this account will lead us to an acceptable reconstruction both of Duhem's and of Poincaré's epistemologies.

As is well known, scientific realism faces a major problem posed by the frequent occurrence of scientific revolutions. Even though there might be—between consecutive theories—a *syntactic* continuity guaranteed by the Correspondence Principle, the *semantic* level displays a chaotic series of upheavals and downright eliminations. The referents of the latest hypothesis often oust those of the old one with which they anyway seem to have little in common. For example: the phlogiston was abolished by Lavoisier; the caloric was replaced by matter-in-motion; Fresnel's mechanical medium was supplanted by Maxwell's electromagnetic ether, which in turn gave way to Einstein's immaterial field; particles were supposed to be reduced to fields, which may however turn out to consist of probability waves; and so on. In the face of such repeated overthrows, how can scientists legitimately claim to be gradually homing in on the real structure of the universe?

This undeniable ontological discontinuity threatens to induce instability at the strictly scientific level, whereas the latter ought, if properly understood, to exhibit a near-cumulative pattern of growth (Duhem 1914, Chapters 2 and 3). This led Duhem to distinguish between the representative part of a theory, an abstract or structural component which he accepted, and an explanatory or interpretative part which he totally rejected. The representative part, henceforth referred to as REP, consists of purely formal relations whose only function is to entail well-tested experimental laws. As for the explanatory component, henceforth denoted by EXP, it claims to provide a realist interpretation of the whole system—it purports to anchor REP in a hidden reality whose existence is warranted by some pet metaphysical doctrine.

EXP suffers from two major drawbacks. On the one hand, far from yielding any new predictions, EXP often proves hardly compatible with REP which does all the empirical work on its own. On the other hand, because of its semantic pretensions, EXP misrepresents the history of science as a series of violent revolutions. Axing EXP thus represents a double gain. First, there is an increase in economy of thought without any loss of empirical content. Secondly, the development of science can now be properly seen as a gradual process over which, thanks to the Correspondence Principle, the mathematical form of physical laws is largely preserved. This sustains our belief that we might be moving towards a 'natural classification', towards a unified system whose internal structure mirrors the ontological realm without having to signify its constituents term-by-term.

According to Duhem, there is a good indication that science is coming closer to a natural classification whenever scientists have at their disposal unified and empirically confirmed systems—more precisely: whenever their hypotheses display a high degree of compactness together with a capacity for yielding hitherto unknown experimental laws, where the latter are subsequently confirmed. This judgement clearly rests on the no-miracles argument put forward by Poincaré in support of structural realism (see below, 2.4).

Though by and large correct, Duhem's position seems to me mistaken when it characterises EXP as *interpretative:* it is as though, according to Duhem, every realist claimed to possess, over and above H, some independent entry into the world of the external objects denoted by the theoretical terms of H. This is unfortunately what most interpretations of formal systems *seem* to presuppose. Within science, there is admittedly a sphere in which we have a direct access to the referents of our terms, namely the observational sphere (see above, 1.7). We have already seen that in testing a hypothesis, we compare a phenomenologically reduced proposition, be it about sense-data, intentional objects, or our own mental states, with a state-of-afairs known to us by acquaintance. Observation reports describe referents to which we have unmediated access. By comparing a basic statement with its intended immanent denotation, we can ascertain a truth-correspondence between the former and the latter. As already explained, it can be claimed that Poincaré regarded these immanent states-of-affairs as the crude and indubitably known facts which science must save at all cost. As for the objects belonging to the external world, they are known exclusively by description, that is: by means of theory alone. This in no way entails that Poincaré looked upon the notion of a correspondence between a theoretical proposition and some transcendent state-of-affairs as untenable; he simply held that such a correspondence cannot be directly ascertained, for we have here a single entry: the theoretical one.

If, as Duhem (wrongly) assumed, the realist insisted on providing a theory H with a strictly *interpretative* component EXP, then such a realist must succumb to one of the following two delusions: a requirement of picturability or an absolute demand for the reduction of every hypothesis to some familiar and allegedly intelligible theory. The picturability requirement, which was explicitly rejected by Poincaré, stems from our wish to have all the objects postulated by theory resemble the intentional entities familiarly known to us by acquaintance. That the latter can be known in this fashion is hardly suprising, for they are engendered by our mental acts; from which there unfortunately follows the desire to picture, say, the electron as a uniformly charged sphere rotating in space. If we are to believe quantum physics, this is however only a mental image.

At this point, it proves useful to introduce a distinction somewhat similar to the one made by Arthur Fine in his 'Shaky Game' (Fine 1986, p. 146). We shall say that a transcendent entity is detectable if its presence can be *causally linked* to our perceptual experience. It is said to be obsevable if its properties are moreover faithfully *represented in* experience. As pointed out by Brentano and Husserl, all acts of consciousness and hence all perceptual experiences are intentional: they constitute and simultaneously aim at so-called intentional objects; the latter are mentally inexistent and need have no real physical entities corresponding to them (see Brentano 1924, Book 2, Chapter 1; also Husserl 1928, Chapter 2). According to the above definitions, the observation, as distinct from the mere detection, of a physical object takes place only when the latter exists and resembles its intentional counterpart (see above, Section 1.8). My view is that no physical object can, in this precise sense, ever be observed. Not only does quantum mechanics tell us that the particles constituting a macroscopic body possess properties absent from all perceptual acts of consciousness; these properties are moreover at odds with the actually perceived qualities of the intentional object. For example, the latter is seen to have a continuous and continuously coloured surface, to possess a well-defined spatio-temporal position, and so forth; these attributes are expressly denied by quantum theory to the ultimate constituents of the macroscopic object. It is furthermore uncertain whether quantum physics enables us to integrate this object out of its parts. Thus no physical entity is ever strictly observed; it is only detected via some intentional counterpart, that is: by means of perceptual acts of consciousness which are not cognitively but causally linked to some physical processes. Even classical physics ascribed to macroscopic bodies properties which are not replicated in perception: although a solid body was taken to consist of point masses separated by finite distances, it is actually seen as continuously spread out in space; and its real shape, its essential primary quality, is not visually given in any direct way—a square object is seen as a rhombus, a circle as an ellipse, and so on. Finally, intentional objects display secondary properties assumed to be absent from their physical causes. All one could claim was that certain properties of perceptual objects vaguely resemble the primary qualities of their external correlates; so Descartes rightly maintained that our knowledge of the physical world is achieved by the intellect and not through the senses. Even if classical physics had been true, we should still have had to speak, not of the strict observation but of the detection of physical entities.

Though presented as a realist desideratum, picturability is in effect an anthropomorphic and even an idealist requirement; for it seeks to assimilate all physical objects to certain intentional entites and hence to mere percepts. Like Planck, who was rightly to maintain that anthropomophism and objectivity pull in opposite directions, Poincaré had put forward the view that:

> Such a philosopher claims that all physics can be explained by the mutual impact of atoms. If he simply means that the same relations obtain between physical phenomena as between the mutual impact of a large number of billiard balls—well and good! This is verifiable and perhaps true. But he means something more, and we think we understand him because we think we know what an impact is. Why?. Simply because we have often watched a game of billiards. Are we to understand that God experiences the same sensations in the contemplation of his work that we do in watching a game of billiards? If it is not our intention to give his assertion this fantastic meaning, and if we do not wish to give it the more restricted meaning I have already mentioned, which is the sound meaning, then it has no meaning at all. Hypotheses of this kind have therefore only a metaphorical sense. (Poincaré 1902, pp. 163–64; see also Planck 1965b)

This passage hints at the second illusion to which the realist, as portrayed by Duhem, might succumb. This illusion stems from the false belief that we can directly intuit the objects postulated by some familiar theory; for the latter appears so intelligible that we imagine ourselves capable of seeing and touching the entities it describes. This is why we are, according to Duhem, misled into requiring that every new hypothesis be 'interpretable' in terms of such entities. This Duhemian analysis is to my mind mistaken; for although we might use the semantic concept of *interpretability*, what we often require is the *reduction* of the new hypothesis to some old theory like mechanics. But mechanical processes, no matter how seemingly familiar, remain transcendent and hence grasped by description, that is: by means of syntax alone; for only theory can tell us what mechanical entities are like.

The illusion that we can directly apprehend certain kinds of external objects is also induced by the terminology used in some mathematical disciplines, for instance in model theory. We are so familiar with Euclidean figures that when we come to interpret spherical geometry in Euclidean terms, we have the impression of associating with certain abstract concepts, such as that of a geodesic, concrete perceptual figures like the arcs of a big circle on a Euclidean sphere. We thereby forget that we are merely translating one formal system, A, into another system, B, in such a way that all the axioms of A become theorems of B. Even though our metatheory might be informal, nowhere do we need to consider anything beyond the formal syntaxes of A and of B. Whenever we interpret a first-order language, we similarly seem to associate with empty signs, for example with names or with predicate letters, concrete objects like the elements of some domain or sets of ordered n-tuples of such elements. We again have the impression of going over from an abstract system to something far more tangible, to something we could almost grasp with our hands. In reality, we are merely using one formal system, namely set theory, in order to determine entities satisfying the axioms of another system. At no point do we really abandon the syntactical approach except at the meta-level; and anyway, the metatheory itself deals with the syntaxes of the formal systems, not with any geometrical objects or with any concrete collections.

Going back to Duhem, the explanatory part of a hypothesis H does not constitute an *interpretation* of H in terms of picturable entities, but an attempted *reduction* of H to some background theory which is judged to be highly intelligible. Duhem rightly remarked that it would be wrong to regard the possibility of this kind of reduction as an adequacy requirement to be imposed on every new hypothesis. But whenever it can be successfully carried out, such a reduction ought surely to be welcomed. For almost by definition and *provided* the observational vocabulary be left intact by the reducing translation, a reduction never diminishes the empirical content of the reduced theory. (This condition of empirical semantic invariance becomes self-evident as soon as we realize that all observational predicates are phenomenological and hence possess fixed meanings. See above, 1.3.) The empirical content may in fact increase if the reducing theory is not a conservative extension of the reduced one. Though aware of the existence of many attempted reductions—usually to mechanistic atomism—Duhem frequently chose to conflate 'interpretation' and 'reduction'. He thereby found it easy to show that a system's explanatory component, which he first took to be strictly *interpretative*, does no empirical work at all, for like all logical consequences, the observational consequences of a hypothesis are not altered by changes in the meanings of any terms not occurring in them; they are consequently independent of any meanings attributed to the theoretical terms. The interpretation of any high-level theory is thus easily shown to be *empirically* otiose. From this, Duhem wrongly inferred that the explanatory component of a system, now understood as the *reductive* part, is also empirically idle.

The above is admittedly a reconstruction of Duhem's position. As already mentioned, he was aware of the existence of many attempted explanations which were clearly meant to be reductions to mechanics. And he went out of his way to show that far from being otiose, most of these reductions are positively harmful. This is because the reducing and the reduced hypotheses are often *made compatible* by means of adhoc manoeuvres. For example: Descartes's law of refraction contradicts his mechanics, which had to be modified in order to acommodate his optics; Maxwell's field equations contradicted his ether hypothesis, which was finally passed over in silence, and so forth. Note that Duhem does not object to realism in its Aristotelian form, but only to *reductive* realism which, when realistically interpreted as atomism, appears to threaten Catholic dogma. His objections are however so global that they hit all forms of reductive explanation; so he throws the baby out with the bathwater. His antipathy to all versions of dynamism moreover accounts for his wrong-headed appraisal of many programmes which were destined to score great successes in the twentieth century, namely: atomism, Faraday's and Maxwell's field theories, Darwinism, and the relativistic hypotheses of Lorentz and Einstein. In his excellent paper on Duhem, Alain Boyer presents a detailed analysis of the subtle but crucial difference between the two notions of realist explanation and of natural classification (Boyer 1992). Boyer shows that no matter how beefed up in a realist way, a classification would almost inevitably lead its author to regard a programme like that of phenomenological thermodynamics as paradigmatic for the whole of science. It should therefore come as no suprise that Duhem rejected the works of Maxwell, Boltzmann, and Einstein.

To sum up, according to Duhem, all explanations, when understood as interpretations, are otiose, and when understood as reductions, *can* be harmful. So he invalidly concluded that all explanations are at best superfluous. As demonstrated above, an extra-linguistic interpretation of theoretical terms can be effected only by means of picturable entities or, as Duhen himself put it, by means of sensuously given objects; and the development of physics has, since the begining of the scientific revolution, gradually discredited this picturability requirement. But explanations qua reductions are a different matter altogether. No attempted reduction *can* do any harm: either it fails, in which case we have to content ourselves with the theory we started from; or if it suceeds, then it never reduces, but could conceivably extend the empirical content of our initial hypothesis. In this latter case, the extra testability of the reduced theory may lead either to new dramatic confirmations—possibly in the form of verified corrections of old predictions—or else to a refutation. In the latter subcase, we can again fall back on the initial hypothesis; but serious doubts should then arise about the tenability of the reducing theory which would have been falsified *modulo* the reducing translation. This is what eventually happened to mechanics qua reducing theory of electromagnetism. It should finally be kept in mind that the translation of one scientific system into another is often approximate and therefore corrective; whence the possibility of crucial experiments between the reduced hypothesis and its old version.

Another incoherence in Duhem's position stems from his thesis that observation statements are theory-laden, that they carry theory-dependent meanings and are therefore fallible. This view is prima facie similar to Popper's (Popper 1959, Chapter 5). But whereas Popper holds high-level theories to be meaningful propositions susceptible of having a truth-value, Duhem regards them as mere schemata for classifying experimental laws; and he fails to tell us how such schemata inject meaning into their obsevational consequences. Of course, Duhem's fideist motives are clear enough: he wants to introduce fallibilism both at the highest (theoretical) and at the lowest (empirical) levels of scientific enquiry. Unlike Popper for whom science is common sense writ large, Duhem accepts the infallibility of common-sense propositions but banishes them from the domain of science. He claims that in passing from common sense to scientific factual statements, we gain precision at the cost of certainty; and we need a Pascalian 'esprit de finesse', some sort of scientific intuition, in order to decide what hypotheses provisionally to accept and what observation-statements to regard as true. Such thoroughgoing fallibilism provides Duhem with an opportunity for regarding the physicists' ability to get closer to a natural classification as a divine miracle. As often happens, this sort of appeal to intuition indicates the absence of any satisfactory solution to the problem at hand, while the underlying motive is to clip the wings of science in order to leave room for faith.

We have already mentioned that, unlike Duhem, Poincaré had no hidden religious or metaphysical agenda. He took factual propositions to possess truth-values which can be infallibly known and hence known independently of any *scientific* conjecture. Obsevational predicates must therefore have meanings fixed in advance of theory. This is the essence of Poincaré's rebuttal of Le Roy's Bergsonian conventionalism: we may have some leeway in inventing theories but we certainly cannot create scientific facts at will. It remains however unclear whether Poincaré took crude factual statements to be about external or about subjective states-of-affairs (see above, 1.3–1.7). We have seen that a rational reconstruction based on his writings yields factual statements as experiential descriptions of the observers' mental states, more particularly of his thoughts and sense-impressions. These reports constitute the phenomenologically reduced empirical basis of the sciences. Thus there is no question, in Poincaré's mind, either of any genuine *interpretation* or of any *direct access* to the referents of theoretical terms—hence his assertion that we may come to know, not the deep nature of such referents, but relations amongst them which are conveyed by theory alone. Our knowledge of the external world is therefore coextensive with that of the syntax, that is: of the mathematical structure of empirically successful theories. In other words, the syntax of unified theories which predict novel facts or establish unsuspected connections between disparate data must be taken to reflect the ontological order of things. Without this assumption we should expose ourselves to the no-miracles objection: we should have to accept the highly unlikely thesis that unified theories happen, purely by chance, to get the facts systematically right (see below, Section 2.4).

As seen above, the demand for an interpretation of scientific laws is bound to lead either to a picturability or else to a reducibility requirement. In the passage quoted on p. 45, Poincaré dismisses the first requirement out of hand. As for the second, let us note that he is not in principle opposed to reductive strategies. A reduction normally diminishes the number of independent assumptions and may thus constitute a step towards the kind of unity which Poincaré highly prized. He nevertheless added two provisos.

First, on being reduced, a hypothesis must preserve its unity. Hence the reducing translation should not be artificially doctored to yield all the known facts while avoiding the known refutations; for such *post-hoc* adjustments typically destroy the unity of the original hypothesis. As examples, let us cite the vain attempts made in the nineteenth century to derive electromagnetism from mechanics while avoiding the prediction of any longitudinal component of wave propagation. Secondly, any excess empirical content should, whenever and wherever possible, be put to the test. Putting it briefly: the reduced theory must keep its unity subject to consistency with the facts. Concerning mechanism, Poincaré denied it any inherent plausibility, but he conceded that it would receive strong support from any successful reduction of other branches of physics to dynamics.

> For a mechanical explanation to be good, it must be simple; to choose it from among all the explanations that are possible: there must be other reasons than the necessity of making a choice. Well, we have no theory as yet which will satisfy this condition and consequently be of any use. Are we then to complain? That would be to forget the end we seek, which is not mechanism; the true and only aim is unity. (Poincaré 1902, p. 177)

2.3 Continuity of Sense or Continuity of Reference?

On the face of it, Poincaré operates with an experientially interpreted empirical basis coupled with uninterpreted high-level hypotheses, which are construed in a purely syntactical way. Is this not the logical positivist stance initially adopted by the Vienna Circle? And is not syntacticism, no matter how liberal, inconsistent with realism, no matter how mitigated? As already explained, the syntactic continuity effected by the Correspondence Principle appears to be belied by the total referential discontinuity obtaining at the ontological level. Irrespective of whether they really exist or not, the classical and the quantum mechanical 'electrons' constitute vastly different objects: the classical particle consists of a well-defined distribution of charge in space while the quantum-mechanical electron is a kind of point mass without any sharp spatio-temporal location. Similarly, relativistic mass is an inertial coefficient varying with velocity while its Newtonian counterpart represents the amount of substance in a given body. In both cases, the same word, *electron* or *mass*, is used in order to mask an abrupt referential discontinuity. Such discontinuities have frequently occurred in the past and will probably do so in the future; but no matter how moderate or structural, scientific realism presupposes a minimum of substantive—as opposed to merely terminological—gradualism. This is why Laudan and Fine declared realism dead (see Leplin 1984).

Attack being the best form of defence, I shall try to refute the above objections by a counter-offensive. Empiricist—and more particularly Anglo-Saxon—philosophers are distrustful of such ghost-like entities as classes and Fregean senses; they feel more at ease with the concrete referents of names. As explained in 2.1 above, Quine maintains that as long as we quantify only over first-order variables, we remain committed to the existence of individuals but not to that of attributes; where the latter *only appear* to be named by predicate constants. Unfortunately for those who field referential discontinuity against realism, words like 'electron' do not denote concrete objects but the kind of attribute or class which Quine rejects. In the absence of such referents, there seems to be no room left for any talk of referential continuity or discontinuity.

It is of course true that the extensions of the classical and of the quantum-mechanical 'electrons' are disjoint. This is however perfectly compatible with the continuity of connotation or of Fregean sense (see Frege 1986). I take it that the concept of an electron consists of a cluster of predicates. Despite being distinct qua intended objects, the classical and the quantum-mechanical electrons may therefore have several attributes in common, for example: those of possessing rest-mass, electric charge, and the capacity for giving rise to an electromagnetic field. Needless to say, if a predicate occurs in one cluster while its negation is implied by the other, the two extensions become disjoint. And yet, as we move from one hypothesis to the next, it seems reasonable to require that there be some overlap between two corresponding notions which often bear the same name. As just explained, such an overlap is to be sought, not in the extensions, but in the connotations of the two concepts. Continuity can moreover obtain, again in the absence of all extensional overlap, in the following way: some predicates admit of varying degrees of intensity which are signified by numerical indices; two such predicates, with different though nearly equal indices, may occur respectively in the property-clusters of the old and of the new notion. Because of this numerical difference, the extensions of the two notions will be disjoint but their qualitative meanings remain nearly identical. For example, the classical and the relativistic masses of a body have a least one property in common: both are measures of the object's inertial resistance to acceleration from rest. For nonzero velocities, they represent different degrees of laziness, but provided v/c is small, this difference is negligible. (We have here to do with the expression $[m_0/\sqrt{(1-v^2/c^2)}]-m_0$, which tends to zero with v/c. See below, Chapter 4.) Hence the antirealist charge of discontiuity rests on a wrong analysis which focuses on the denotations rather than on the meanings of certain terms. The insistence on referential continuity actually makes as little sense as the requirement that an old hypothesis should not only tend towards, but also be consistent with, the new one.

Prima facie, the demand for some overlap between the extensions of corresponding concepts is analogous to a consistency condition, which seems so minimal as to be eminently reasonable. But this is not how science progresses. Putting it bluntly, inconsistency is perfectly compatible with continuity; for a function is generally different from its limit and a law will usually be inconsistent with all of its limiting cases. We should therefore require that sequences of successful theories display a near-cumulative pattern only with regard to their Fregean senses, while any demand for referential continuity would lead us to adopt a principle of charity according to which Newton, Huygens, Young, Fresnel, Maxwell, and Einstein were, unbeknownst to themselves, talking about the self-same entity, namely the photon. This claim may be considered true in the far-fetched sense that photons were the hidden causes of various observations made by all these scientists; but this principle of charity in no way explains the actual continuity in the development of optics from Newton to Einstein, for successive scientists have conceived of this hidden cause in logically irreconcilable ways. This is why the thesis of connotational continuity seems to me to provide the only solution to this puzzle.

One question still remains unanswered: is there any connection between meaning and syntax? According to Frege, the sense of a well-formed expression is mirrored by its syntactical structure (Frege 1986). For example: unsaturated entities or concepts are signified by incomplete expressions such as F( ), which exhibit gaps. More importantly, the meanings of complex expressions are built from those of their component parts. The syntax of an expression mirrors the structure and complexity, not of its referent, but of its sense. By insisting on syntactical gradualness, Poincaré was in effect imposing a continuity requirement on sense or on structure. Whenever this form of continuity obtained in a sequence of unified and empirically successful theories, he regarded it as showing that science had latched on to real connections between things.

> Thus Fresnel's theory enables us to do today as well as we did before Maxwell's time. The differential equations are always true, they may always be integrated by the same methods and the results of this integration still preserve their value. It cannot be said that this is reducing physical theories to simple practical recipes: these equations express relations and, if the equations remain true, it is because the relations preserve their reality. They teach us now, as they did then, that there is such a relation between this thing and that; only the something which we then called motion, we now call electric current. But these are merely the names of the images we substitute for the real objects which Nature will hide forever from our eyes. The true relations between these real objects are the only reality we can attain, and the sole condition is that the same relations shall exist between these objects as between the images we are forced to put in their place. If the relations are known to us, what does it matter if we think it convenient to replace one image by another? (Poincaré 1902, p. 161)

The above passage was quoted in full because it summarises Poincaré's position concerning the status of scientific theories. Let us note the Kantian flavour of the thesis that the deep nature of things cannot be directly apprehended; for the latter are not grasped as they are in themselves but as they appear through the medium of experience and thought. As is well known, the linchpin of Kant's argument for the inaccessibility of noumena to theoretical reason is the presumed existence both of necessary intuitions and of synthetic a priori principles: prior to experience, we can infallibly anticipate some fundamental aspects of the phenomenal world, for example those features connected with space, time, and causation without which no phenomena could be apprehended by us. Kant concluded that these features do not belong to the noumenal world but are contributed entirely by the transcendental ego, for our knowledge of them would otherwise have to be synthetic a posteriori, which it is not. Poincaré was aware of the invalidity of this inference, or rather of the falsity of its premises. He held that we are capable of no specifically spatial intuition and that geometry was neither analytic nor synthetic a priori, nor even necessarily Euclidian. This partial breakdown of the Kantian system left him free to claim that we can know, if not the intrinsic properties of the noumena, then at least some true relations between them; but since we cannot directly grasp the nature of any things-in-themseves, all picturability requirements are null and void, except qua heuristic devices.

Consistently with the views expressed in the passage quoted on p. 52, we can put forward the following conjecture: Poincaré would have attached little significance to the extensional equivalence of two theories which possess different mathematical structures or else involve relations having very different meanings; he would have regarded the two theories as giving fundamentally different accounts of the physical world; for what mattered most to him were the universals—understood intensionally—and not the objects subsumed by them. His discussion of Newton's, Lagrange's, and Hamilton's contributions to physics shows that he regarded Hamilton's equations as constituting a theory different in kind from that of Newton; for despite being extensionally equivalent under a *synthetic* translation, these two hypotheses have distinct mathematical structures.

> To form these equations we need not know the relations which connect the parameters q with the coordinates of the hypothetical molecules, nor the masses of the molecules,nor the expression of U (= potential energy) as a function of the coordinates of these molecules. All we need to know is the expression of U as a function of the parameters q, and that of T (= kinetic energy) as a function of the parameters q and of their derivatives—i.e. of the expressions of the kinetic and potential energies in terms of experimental data. (Poincaré 1902, p. 220)

Had he lived longer, Poincaré would certainly have felt vindicated by the fact that wave mechanics took its starting point not from Newton's but from Hamilton's equations. This development would have demonstrated that we are dealing here with two vastly different schemes which happened, at one time, to be observationally indistinguishable. Poincaré would have similarly looked upon Schrödinger's wave-theoretical approach as being distinct, and as suggesting a fundamentally different heuristic from

Heisenberg's matrix mechanics. This would again have been because he viewed the mathematical structure of a theory as representing its most realist aspect.

To sum up, whereas referential continuity demands that as we move from one hypothesis to the next, we continue to talk about *roughly the same objects*, structural continuity more reasonably requires that we talk, in *similar terms*, about *one* thing-in-itself which is not directly accessible to us. Even this last formulation seems to me to be too stringent; for we ought only to require that the meanings of whole propositions, as distinct from those of their separate terms, should be approximately preserved across revolutions. In other words: there ought to be a translation of the old system into the new one such that:

(a) all observational functions and predicates remain unchanged, and

(b) the old axioms are transformed into theorems, or else into limiting cases of theorems of the new theory.

Such a liberalised Correspondence Principle may enable us to understand why the phlogiston programme might well have been progressive when it was overtaken by Lavoisier's theory; and this despite the fact that 'phlogiston' has null extention: it fails altogether to refer. All we probably need to do is follow Russell's method of treating certain expressions as incomplete symbols: contexts in which the presence or the loss of phlogiston is asserted should be translated as those in which an absence or a gain of oxygen is respectively registered. Dephlogisticated air will be interpreted as air rich in oxygen, and phlogiston should be regarded as anti-oxygen. Thus, given the logic of their situation, the phlogistonists were ultimately compelled to attribute negative weight to phlogiston; or if they baulked at this idea, they had—reluctantly—to accept Lavoisier's oxygen theory, which many actually chose to do.

2.4 The Structure of Poincaré's Argument in Favour of Structural Realism

As already mentioned, since we cannot and should not, according to Poincaré, attribute the success of the sciences to mere chance, we must accept that at least some of our theories express real connections between things. By 'things' he means something like the Kantian noumena. Thus Poincaré is the real author of the no-miracles or no-conspiracy argument, which he invokes in arguing for the empirical adequacy of well-corroborated laws:

> We have verified a simple law in a considerable number of particular cases. We refuse to admit that this coincidence, so often repeated, is a result of mere chance and we conclude that the law must be true in the general case. Kepler remarks that the positions of a planet observed by Tycho are all on the same ellipse. Not for one moment does he think that, by a singular freak of chance, Tycho had never looked at the heavens except at the very moment when the path of the planet happened to cut that ellipse. What does it matter then if the simplicity be real or if it hide a complex truth? Whether it be due to the influence of great numbers which reduces the individual differences to a level, or to the greatness or the smallness of certain quantities which allow of certain terms to be neglected—in no case is it due to chance. This simplicity, real or apparent, has always a cause. We shall therefore always be able to reason in the same fashion, and if a simple law has been observed in several particular cases, we may legitimately suppose that it will still be true in analogous cases. To refuse to admit this would be to attribute an inadmissible role to chance. (Poincaré 1902, pp. 149–150)

Poincaré was also aware that concurrently with the no-conspiracy argument, he was implicitly addressing the insoluble problem of induction:

> I do not at all wish to investigate here the foundations of the principle of induction; I know very well that I shall not succeed; it is as difficult to justify this principle as to get on without it. I only wish to show how scientists apply it and are forced to apply it. (Poincaré 1906a, p. 133)

He then went on to extend the no-miracles argument to the case of high-level theories as distinct from that of directly refutable generalizations:

> It will be said that science is only a classification and that classification cannot be true, but convenient. But it is true that it is convenient; it is true that it is so not only for me, but for all men; it is true that it will remain convenient for our descendants; it is true finally that this cannot be by chance. (Poincaré 1906a, p. 140)

Remember that Poincaré declares a hypothesis convenient if it is unified and if it explains the phenomena without adhoc adjustments. It can hardly be doubted that in the last sentence of the above passage, he treats the degree of convenience of a theory as an index of its verisimilitude. This fits in well with his repeated claim that through its mathematical structure, a successful system reveals true relations among real objects.

In his *Shaky Game*, Arthur Fine delivered the following powerful criticism of the no-miracles argument. Scientific realism ought to be treated as a scientific hypothesis which bears the same relation to the success of the sciences as does an empirical theory to a set of data; but the antirealist is precisely the kind of person who rejects all abductions, that is: all inferences to the best explanation; so that even if he accepted the success of science as a fact, he would still refuse to infer from it the truth of realism, just as he would reject the move from the truth of ordinary basic statements to that of scientific hypotheses. In Fine's view, the realist is thus caught in a hopeless circle; he is like an inductivist who tries to persuade an anti-inductivist of the truth of some conclusion by means of an inductive inference.

In order to show the invalidity of this alleged confutation of realism, I need once again to introduce distinctions somewhat similar to those made by Fine himself. Realism has often been characterized as the belief in the existence of a reality independent of the human mind. I find this definition unsatisfactory because it presupposes an ontological schism between the mind on the one hand, and that which it tries to grasp on the other. By metaphysical realism, I shall understand the thesis that there exists a structured and undivided reality of which the mind is part; moreover that far from imposing their own order on things, our mental operations are in their turn governed by the fixed laws which describe the workings of Nature. In other words, our brain or mind evolves according to the very laws it sets out to capture. As shown by Planck and by Popper, this thesis entails that a total explanation of everything in the universe can never be achieved by the human mind: on pain of vicious circularity, the brain is in principle unable to predict its own future states. Far from being incompatible with metaphysical realism, this negative result could be regarded as one of its welcome consequences (Planck 1965d; Popper 1982, Chapter 3).

Methodological realism consists of the further assumption that the structure of reality is, at least in part, intelligible to the mind. It is therefore reasonable for scientists to aim at grasping that reality. Of course, they may choose not to do so and content themselves with constructing phenomenological laws.

Henceforth referred to as SR, scientific realism finally adds the thesis that successful theories—those unified systems which explain the data without adhoc assumptions—are approximately true. That is: such systems reflect, in one way or another, the order of things as they are in themselves; successful hypotheses thus model, and in some sense 'correspond' to the real structure of the world. Note that SR is commited neither to the correspondence nor even to Tarski's semantic theory of truth. There is no reason why the way in which a hypothesis mirrors reality should be the usual term-by-term mapping described by traditional semantics. As already explained, the mathematical structure of a scientific system may globally reflect reality without each component of the system referring to a separate item of that reality. This is the structural-realist stance adopted by both Duhem and Poincaré. In its defence, we can say that our usual semantics is inspired by a common-sense ontology consisting of discrete *intentional* objects. Such an ontology need however not be that of the *external* world. In other words, common sense tends to assimilate the transcendent to the intentional sphere, and hence to posit fixed and well-circumscribed entities subsumed by various predicates, but this is not necessarily the way Nature operates. As already mentioned, external objects are known exclusively by description, that is: through the mediation of a network of theories. So we cannot directly ascertain what kind of correspondence might link our successful hypotheses to the transcendent world. Only idealists like Berkeley and Husserl could feel sure about the nature of such a correspondence, for idealism postulates only *immanent entities*. By contrast, the way in which realist theories latch on to the world constitutes a *problem* for which there exist only conjectural solutions. Thus SR should not be made to stand and fall with any specific theory of truth; all it asserts is that highly corroborated theories reflect the ontological order in a way which the scientist may—or may not—choose to specify. There is no reason why realists should not adopt a fallibilist attitude as regards both the truth-value and the mode of reference of their conjectures. Let us not forget that Duhem aptly compared a scientific concept to the flat pieces forming an armour which never exactly fits a knight's body. Such concepts have a symbolic or metaphorical, rather than a strictly referential function. (For the important role which metaphors play in scientific explanations, see MacMullin 1984.) Poincaré's rider should nonetheless never be forgotten: the meanings and referents of the observational or phenomenological terms must be fixed in advance.

Going back to the three positions outlined above, note that each is stronger than the one preceding it. This is illustrated by the following examples: it can be claimed that Kant and Duhem were not methodological but metaphysical realists. As for Poincaré, he firmly subscribed to SR whereas a scientist like Mach rejected realism altogether. These two philosophers thus occupy the opposite ends of the scale. Finally, Popper was a methodological realist but his attitude to SR remained interestingly ambiguous. He initially based his own version of SR on a definition of verisimilitude linked to the correspondence theory of truth, but David Miller undermined this definition by showing that it applies to no false theories at all. Whereas Miller might still accept methodological realism whilst rejecting SR, Popper fell back on an intuitive notion of verisimilitude, thus indicating that he still adhered to SR.

As already mentioned, Fine levels a charge of vicious circularity at the no-miracles argument adduced in support of SR. Like Fine's critisism, my first rebuttal of it will be somewhat scholastic. Put: SR('H') = ('H' is approximately true). Even the most hard-headed instrumentalist must accept some theories, not as being true, but as offering the best account of certain data. Should he regard the empirical success of a given hypothesis H as a datum and should this success be entailed by SR('H'), then the instrumentalist would have to accept, not the truth of 'SR('H')', but SR('H') simpliciter. He would thus have to acknowledge the approximate truth—not of 'SR('H')' itself—but of 'H'. As usual, this separation between different levels avoids all vicious circularity.

The instrumentalist has of course plenty of room left for manoeuvre: he could deny either that the success of H constitutes a 'fact' in the ordinary sense of the word, or that there exists any genuine inferential link between SR('H') and the success of H. Admittedly, the empirical success of H could have been ascertained only in the past and only on finitely many occasions. However, the instrumentalist typically tends to believe in induction and is therefore unlikely to question the kind of reasoning which generalises the confirmation of H to some infinite domain, D, say. Thus the instrumentalist will be inclined to accept the empirical success of H in D as a fact.

As for the putative inferential link between SR and the success of H, let me first remark that Fine's argument is predicated on a contentious assumption, namely that *scientific* realism deserves its name: SR is taken to be an empirical hypothesis whose factual basis consists of the successes of various scientific theories. Against this claim, I propose to show that SR is a metaphysical thesis, which would incidentally imply that any logically weaker position, such as metaphysical or methodological realism, is also metaphysical.

Let us again follow Popper in holding a proposition N to be metaphysical if N is empirically irrefutable or, more generally, if N forms no essential part of any scientific system S; that is, N belongs to no S such that for some observation report e, e follows from S, but not from S without N. (Note that S must first have been axiomatized in a natural way. See below, Appendix 2.) Taken in isolation, SR is trivially irrefutable; for in order to falsify it, we would have to produce a theory H such that:

(a) H is unified, highly empirically confirmed, and

(b) H is not even approximately true.

We have discussed some difficulties pertaining to (a). More importantly however, (b) is totally empirically undecidable; for we have at our disposal no theory which is both *known* to be true and with which H could be effectively compared. In support of their position, antirealists have argued that Newton's theory, though vastly different from Einstein's, remains empirically successful in many domains. So don't the Newtonian laws constitute the kind of highly confirmed hypothesis which cannot even be approximately true? One possible answer to this objection is that Einstein's theory is not *known* to be true; not being quantized, it is in fact almost certainly false. Another reply is that the domains in which the Newtonian laws are successful are precisely those where Newton's equations form limiting cases of Einstein's. Thus even if Einstein's theory is taken to be approximately true, then so is Newton's—at least in the considered domains. Let us incidentally note that SR does not entail the success of any specific theory H: in order for SR to do this, the approximate truth of H must first be reliably known; and such knowledge has been shown to be unattainable. *Given* the empirical success of H, SR would however enable us to draw an *empirically undecidable* conclusion about the approximate truth of H. (Throughout this argument, I have implicitly assumed that the approximate truth of H, when restricted to some domain D, entails the reliability of H within D; which appears to me unproblematic).

We still have to examine whether SR can conceivably figure as an indispensable premise in the derivation of some empirical prediction, e. This is highly unlikely. The only way in which SR could perform such a task is through SR('H'), that is, through the assertion that some scientific theory 'H' is approximately true. But this assumption is weaker than the truth of 'H', which is in turn equivalent to H. Thus, postulating H renders SR('H') superfluous. All this goes to show that within a *narrow definition* of scientific practice, a redundancy theory of truth can be consistently adopted. Had Fine contented himself with making this point, then his position would have been impregnable; but he also wanted to forbid scientists and philosophers, when *reflecting* on their trade, to espouse any theory of truth whatsoever. More on this later (Section 2.5).

It follows from the above that SR and hence all three forms of realism are metaphysical theses; so the whole basis for Fine's argument collapses. But does this also mean that SR cannot be argued for at all? Far from it, as the following considerations will, I hope, show.

Looking more closely at Duhem's and Poincaré's arguments in support of SR, we see that they all turn on the cogency of the following inference. Recall that for Poincaré, the principle of unity is a transcendental condition which must be imposed on all scientific hypotheses. So let H be a unified system, one whose various parts are closely knit together, and let D be the empirical domain of H. Should one part of H diverge considerably from the truth, then thanks to the the unity of H, the falsity of this one component will have repercussions throughout the system, and more particularly in those parts which govern empirical prognoses. H can therefore be expected to get the facts wrong in whole subdomains of D. Suppose however that H has actually been confirmed whenever and wherever it was tested. Only finitely many tests could admittedly have been conducted and even a hopelessly false theory may get some of the facts right. But then we would be compelled to assume that an evil demon made us test H only at those points where H happened to make correct predictions. Otherwise, the chances are that we should have stumbled on one of the many falsifiers of H. (See Poincaré's imaginary example about Tycho observing Mars only when Mars happens to lie on an ellipse.) Poincaré argued that this kind of conspiracy is highly unlikely since it relies on the existence either of an evil demon or of an improbable sequence of coincidences. Note that such a conspiracy theory is on a par with the counterinductive proposition that high corroboration is divorced from future reliability. It is therefore likely that H is at least approximately true.

No matter what one thinks of this intuitive anti-conspiracy argument, it remains a central tenet of scientific practice. For example, Poincaré could not bring himself to believe that only the first- and second-order effects of the earth's motion systematically failed to materialize; so he concluded that relativity is a universally valid principle. Maxwell similarly refused to accept that light should have exactly the same velocity as all electromagnetic disturbances and yet differ fundamentally from the latter. Finally, Einstein could not believe that despite being caused by absolute velocities, electromagnetic processes systematically give rise to phenomena dependent only on relative motions, and he found it odd that despite representing two fundamentally different properties of matter, gravity and inertia should always remain in unison at the observable level. To repeat: in all these cases, scientists refuse to believe that some Cartesian demon set out to deceive them about what really happens behind the scenes.

The defence of SR thus rests on principles like that of the rejection of conspiracy theories which, far from being imported from outside science, lie at the heart of scientific activity. This is why I cannot agree with Fine when he maintains that:

> realism requires substantially much more than just the truth of certain claims. What it requires is that this truth (or these claims) have a certain special aura—so to speak—an aura which reaches out to reality itself. Since this realist emanation is not to be found in the mere fact of the truth of these existence claims, I have asked

> whether it might be found to be attatched to those claims by the scientific practices that ground their truth. But although those practices are somewhat elusive, they are still definite enough to be seen not to involve the subtle emanations of realism. I conclude that when the realist attaches the aura of his realism to science, he goes well beyond what is warranted either by the truth of specific scientific claims or by the features of scientific practice that ground those claims. . . . The 'externality', the reaching out to definite features of a definite world-structure, is not to be found in science itself. It is a kind of overlay to the text of science. (Fine 1986, p. 154)

By granting the truth of specific existence claims, Fine actually concedes to structural realism more than the latter really bargains for. As for Fine's negative theses, I have shown that SR is deeply rooted in scientific practice: it relies on the same argument which enables both theoreticians and technologists to move inductively from the finite corroboration of a law to its general empirical adequacy.

2.5 Does Realism Have a Cognitive or Merely a Motivational Content?

One might try to adopt the minimalist attitude advocated by Fine under the name of NOA (Natural Ontological Attitude; see Fine 1986, Chapter 7). This seems to consist of the following injunctions: be neither a realist nor an antirealist so that you will have no need of any notion of approximate truth; if you believe in a theory, then you ought to be committed to the existence of all referents of its terms and to the truth of all its logical consequences; above all, do not go beyond the limits prescribed by scientific practice, for doing so would hamper scientific progress.

It seems to me that Fine condemns physicists and philosophers either to an amnesiac or to a schizophrenic existence, or else to believe neither in the strict nor even in the approximate truth of any theory whatsoever and hence to be antirealists after all. For if a scientist has any knowledge of the history of his subject, he will realize that his predecessors believed in a series of hypotheses each one of which was rejected in favour of the next. Unless he forgets about covariance, the physicist is moreover unable to believe in the truth of nonrelativistic quantum mechanics, and unless he forgets about quantization, he cannot believe in the existence of classical relativistic fields. Given that Quantum Field Theory has not yet assumed its final form, then not being allowed any notion of approximate truth, the scientist can believe in *no* statement transcending the barest observational level; he is reduced to using 'models' merely in order to obtain empirical results. He might just as well be an honest instrumentalist or—at best—a conventionalist. Instrumentalism is a tenable position, which is however rejected by Fine on the grounds of its antirealist and hence nonminimalist character. It could of course be maintained that NOA describes a methodological attitude which has nothing to do with actual history. But one way—proposed by Lakatos—of testing a methodology M is to attribute an intuitive version of M to somebody who, by common accord, is regarded as a great scientist; then to examine how many of the scientist's actions can be explained by M without recourse to external factors for which there exists no independent evidence. (Note: 'external' means 'lying outside M'.) In fact, Fine himself applies this method to Einstein's case. Having imputed to Einstein an intuitive adherence to NOA, Fine faced the formidable problem of accounting for Einstein's strong pronouncements in support of SR. To my mind, this is how Fine finds himself compelled to interpret Einstein's realism as constituting, not a cognitive, but a purely motivational stance:

> My argument then is that if we understand Einstein in the way he asks us to, his own realist sounding language maps out a position closer to constructive empiricism than to either 'metaphysical realism' or 'scientific realism'. . . . For, if we look again at his Columbia address, then we see that whereas he quickly backs off the cognitive force of his remarks, he fully embraces the motivational force of that realism—especially by contrast with the programme of positivism. (Fine 1986, pp. 108–09)

And:

> The particular psychoanalytic concept that seems to me most apt for Einstein's realism is the concept of an imago—the complex ideal of the parent, rooted in the unconscious, elaborated by means of childhood fantasies, and bound with the strong affect of that childhood period. The parental imago lies behind and drives certain of our behaviour just as Einstein tells us that realism 'lurks behind' and drives our work as scientists. . . . In Einstein, I would suggest, the fantasy elements associated with his realist imago were expressed precisely by his realist vocabulary. . . . I think Einstein knew perfectly well that spinning the tales of 'reality' was just letting childhood fantasies have their head. (Fine 1986, p. 111)

I have to agree with Fine that whenever they read passages like the above, realists of my ilk feel an irresistible urge to stamp their feet or to reach for their guns. Not that psychoanalytic speculation is per se absurd or unilluminating, but it should not be invoked solely in order to rescue an underlying methodology like NOA. Externalist explanations must be supported by independent evidence. Fine is aware of this requirement, but the only evidence he adduces for Einstein's exclusively motivational realism is that Einstein often baulked at making any direct mention of the actual truth of scientific propositions:

> It is difficult even to attatch a precise meaning to the term'scientific truth'. Thus the meaning of the word 'truth' varies according to whether we deal with the facts of experience, a mathematical proposition or a scientific theory. 'Religious truth' conveys nothing clear to me at all. (Einstein, as quoted by Fine 1986, p. 90)

Note first that Einstein is concerned mainly with the *precision* of the truth-concept; secondly, that he contrasts the notion of religious truth with all others. It seems to me that he held only religious truth to have an exclusively motivational significance. There are also good *independent* reasons why, as with most of his contemporaries, he should have been hesitant about using any explicit concept of truth. There was, at the time, a strong suggestion that truth is intrinsically indefinable. Even a thoroughgoing Platonist like Frege held that despite constituting the core concept of logic, truth could not be defined without vicious circularity ('p' is *true* iff it is *true* that 'p' corresponds to the facts). Popper, who was to become the most ardent defender of realism, wrote his *LSD* without making use of any truth-concept. Russell was a rare exception in that his theory of logical types enabled him to put forward as many types of truth-predicates as there are different orders of sentences. It is hardly suprisingly that in his Introduction to Wittgenstein's *Tractatus*, Russell should have proposed the distinction between object- and meta-language which Tarski later applied to the construction of his truth-definition (see below, 5.4). The philosophical significance of Tarski's definition was that it made all explicit references to the notion of truth intellectually respectable. As far as I know, Einstein was not familiar with Tarski's technical work; so the above list of thinkers adequately demonstrates why many scientists, and Einstein among them, avoided referring directly to the truth of scientific hypotheses. When plying their trade, physicists did not anyway need to mention explicitly any specific notion of truth. The following passage however shows that far from being a NOA'er, Einstein subscribed to a structural form of SR with very high cognitive import:

> If it is true that the axiomatic foundations of theoretical physics cannot be derived from experience but have to be freely invented, can we at all hope to find the right way? Or worse still: is this 'right way' only an illusion? Can we hope to be safely

> guided by experience when there exist theories like classical mechanics, which are largely empirically adequate but do not capture the deep nature of things? To this I answer, with complete confidence, that this right way exists and that we are capable of finding it. In view of our experience so far, we are justified in feeling that Nature is the realisation of what is mathematically simplest. It is my conviction that we are capable, through pure mathematical constructions, of finding those concepts and the law-like connections between them which yield the key to the understanding of natural phenomena. Useful mathematical concepts can be suggested by, but can in no way be derived from, experience. Naturally, experience remains the only criterion for the usefulness of mathematical constructions in physics. But the really creative principle lies in mathematics. In a certain sense, I consider it therefore to be true—as was the Ancients' dream—that pure thought is capable of grasping reality. (Einstein 1934a, p. 116; my translation)

I do not see how such passages can be reinterpreted either in psychoanalytic or even in purely motivational terms. Einstein clearly subscribed to a structural version of realism with powerful heuristic and cognitive content: the mathematical form of a successful theory models and hence reveals the structure of a universe which remains largely inaccessible to our senses. By speculating about the mathematical structure of the laws of Nature, we might therefore be able to devise fruitful heuristic strategies; these may consist in the systematic application of metaprinciples like those of relativity, of least action, and of correspondence. Precisely this had been Poincaré's message.

3

Poincaré's Philosophy of Geometry

Geometry has pride of place in Poincaré's philosophy of science. The received view is that his 'conventionalism' can be traced back to his work on non-Euclidean geometries: through constructing models of any one geometry within any other, Poincaré was allegedly led to the view that the choice of a geometry, as well as of certain basic scientific postulates, is a matter of arbitrary convention. In this chapter, I propose to show that the expression 'geometric conventionalism' already constitutes a misnomer, that in conformity with his global philosophy of science, Poincaré adopted a realist, albeit a fundamentally fallibilist, attitude regarding the status of physical geometry.

3.1 Conventionalism and Geometry

Were it not for my Popperian qualms about asking 'What is' questions, I should have entitled this chapter 'What is Geometrical Conventionalism?' I shall in fact be inquiring into the various meanings of the word 'conventionalism', but only with a view to answering the following question: are there aspects of our knowledge which depend solely on arbitrary decisions and can therefore refer to no external reality?

In one sense, the answer is trivially affirmative: in deciding to formulate his hypothesis in one natural language rather than in another, the scientist is guided, not by epistemological, but by pragmatic considerations such as: should I write my articles in English and thus be understood by a wide public, or should I use my mother tongue in which I feel more at ease? Note that I am here talking about science, not about philosophy. Some philosophers are known to have claimed that deep *philosophical insights* can be rendered only by very special languages, such as Greek and German (see for example Heidegger 1953). I take it for granted that *scientific thought* is equivalently expressible in all natural languages. We shall see that the trivial answer just given was taken to subsume the case of geometry: the choice between Euclidean and hyperbolic geometry was seriously held to be on a par with that between French and German. But more of this later.

At this point, the reader ought to be warned that my approach will be historical, or rather quasi-historical in a very specific sense. Many of the problems to be tackled are admittedly atemporal, but for reasons which are not altogether clear, a historical account makes certain theories far more intelligible than does a hypothetico-deductivist presentation. By way of illustration, we can cite the deep understanding Einstein gained from reading Mach's *Mechanics*, whose approach is both philosophical and overtly historical (Schilpp 1949, p. 20). This intimate connection between the history and the philosophy of science stems from the fact that history proceeds *analytically*, from problems to proposed solutions and then on to further problems, and so on indefinitely. In other words, history describes the heuristic path leading to the construction of scientific hypotheses and thereby explains why the solution to a given problem took one particular form rather than another. As for the synthetic approach, it *starts* from a global system of which various problems and their solutions provide mere illustrations. And one is expected to understand this commanding position through mere proof-checking: by going laboriously through long chains of reasoning, one verifies that answers to various questions are deductively yielded by the system under consideration. In this process, all heuristic insight is lost (see Lakatos 1976, Appendix 2). My intention is not thereby to undermine hypothetico-deductivism, which remains the position most appropriate to *the context of justification*; but on its own, hypothetico-deductivism does not enable us to grasp all the aspects of a philosophical thesis or of an empirical theory. It has of course to be conceded that, in the philosophy of science, one could follow a method which is both ahistorical and problem-orientated; but the most convincing problems remain those which occurred historically and not those artificially tailored to the needs of a scientific or of a methodological position.

There are other reasons for considering history as inseparable from the philosophy of science. As already mentioned, methodologies can be taken to be either supported or else undermined by historical events (see above, 2.5). In so far as they share a number of singular value-judgments about past scientific achievements, two philosophers can engage in a rational debate about the advantages and the demerits of their respective positions. For example, a methodology might be deemed unacceptable if it fails to characterize Newton's, Maxwell's, Einstein's, and Schrödinger's scientific contributions as great leaps forward. It is naturally assumed that descriptions of historical facts can be made, if not value-free, then at least value-neutral with respect to competing methodologies. Thus value-laden concepts like 'progress', 'adhoc', and so forth should initially be kept out of all historiographical accounts.

History moreover plays an important role in the appraisal of scientific laws and hence in the context, not only of discovery, but also of justification. As already explained, a theory should be regarded as *strongly supported* only by those facts to which it was not adjusted in advance. (Should it have been so adjusted, then it might still receive some—*reduced*—measure of support from the facts in question. See above, Chapter 1). The method of construction of a law and, by the same token, its historic past may therefore play an important role in its methodological assessment. For example, Michelson's results confirm Lorentz's and Einstein's theories more strongly than they do Fitzgerald's contraction hypothesis; and the Balmer series, though known in advance of Bohr's atomic models, lends strong support to the latter; for as a matter of historic fact, Bohr did not use the Balmer series in constructing his laws (see Zahar 1989, pp. 11–17). Let us note that a leading figure of the Vienna Circle, namely Moritz Schlick, seems to have shared this methodological viewpoint:

> With the help of this hypothesis, he [Seeliger] actually succeeds in maintaining without contradiction this idea of a world infinitely extended, filling all space with matter of a mean density. An unsatisfactory feature of this theory is, however, contained in the fact that the hypothesis is invented *adhoc,* and is not occasioned or supported by any other experience. (Schlick 1920, p. 70)

Finally, history enables us to demarcate between the internal and the external factors which might have contributed to the form and content of certain theories. Using history, we can decide whether some component of a scientific system arose from problems posed within its domain of discourse, or from its anticipated moral and political impact, or even from considerations of its social relevance. For example, whereas an accident of birth made Poincaré write up his results in French rather than in German, strictly internal considerations arising from his work on Fuchsian functions led him to his so-called 'conventionalism' (see below, 3.6).

Having shown that history is relevant to heuristics and hence to the intelligibility of all theories; that it plays a role in the appraisal not only of methodologies but also of scientific laws; and finally that history enables us to differentiate between the internal and the external aspects of physical hypotheses, let us now consider Poincaré's theory of physical geometry.

3.2 Poincaré's Concept of Physical Geometry

Kant maintained that geometry is about the outer form of our sensibility: in order to experience certain (external) phenomena, we have to imbed them in subjective space whose properties are necessarily governed by Euclidean geometry. This explains why we have, prior to experience, the certainty that all perceptual objects must conform to Euclid's laws (Kant 1953, pp. 36–41). Geometry is thus necessarily Euclidean and it systematically applies to all outer phenomena, whence its empirical untestability.

In 'Geometry and Experience' Einstein claimed that as a branch of pure mathematics, geometry has nothing to do with experience and is subject exclusively to a consistency condition. Through physically interpreting its primitive terms, we can however turn geometry into a science open to empirical falsification; Euclidean geometry is in fact one of the earliest physical hypotheses (Einstein 1934b). As for Poincaré, he held a prima facie paradoxical position lying somewhere between those of Kant and Einstein. Unlike Kant, he knew that 'deviant' geometries could not only be consistently defined but also be applied to physics. However, unlike Einstein, he added that none of these geometries is experimentally refutable.

It is well-known that a great achievement of nineteenth-century mathematicians was to have demonstrated the relative consistency of non-Euclidean geometries, their equi-consistency with the Euclidean system. Kant's contention that the mind is born equipped with a three-dimensional Euclidean continuum had therefore to be either abandoned or else considerably modified. Poincaré's project can be seen as an attempt to recast this important aspect of Kantian philosophy. Poincaré distinguishes between two kinds of space: the mathematical variety, and the representative one. The latter is constructed out of sense-data through a process presupposing no notion of space whatever. What is needed is only our ability to remember sensations, to order them in time and to decide whether or not two past sensations are discernible. (Let us incidentally note the ontological primacy of time over space.) Poincaré goes on to show that the mathematical properties of geometric space, such as dimension and group-structure, possess representative counterparts. This is done as follows. All changes in our sensations are divided into two categories:

> (a) the internal changes which are voluntary and accompanied by muscular sensations,
> (b) the external changes having the opposite characteristics. (Poincaré 1906a, p. 48)

Note that in the language of realism, external changes are alterations perceived to be taking place in the world outside us; whereas internal changes are the muscular sensations associated with the voluntary movements of our own bodies. Moreover:

> We then observe that among the external changes are some which can be corrected thanks to an internal change which brings everything back to the primitive state; others cannot be corrected in this way. Thence comes a new distinction among external changes: those which may be so corrected we call changesof position, and the others changes of state. (Poincaré 1906a, p. 48)

Let us give an example. Consider an object, α, which is displaced from position A to position B. This will cause some change in my visual perception. Should I however move my body so as to re-occupy the same position relatively to α, then I shall restore my initial perception of this object. Thus an external change—the alteration in my visual field as α moves from A to B—is corrected by an internal one, namely the muscular sensations accompanying my movements.

Poincaré then divides the set of internal changes into equivalence classes, where any two members of the same class correct the same external changes. These equivalence classes constitute the group of displacements within representative space. Poincaré moreover holds it to be impossible for us to picture empty representative space. Thus a spatial point is identified with some material object, or rather: with the tactile sensations produced by the contact of the object with a given finger. Amazing as it may seem, starting from this scanty material, Poincaré manages to show that our tactile representative space forms a three-dimensional physical continuum. Visual space is similarly constructed. It consists, first of all, of a two-dimensional image on the retina; the third dimension is then provided either by the convergence of the two eye-axes or by the movement of accommodation of the ciliary muscle. It so happens that convergence and accommodation are always in unison. Were it not for this contingent fact, then visual space would have had four dimensions, which establishes that n-dimensionality, where $n \geq 4$, is not only logically possible but also physically representable (Poincaré 1958b, p. 54). Poincaré adds that if the group laws of displacements were not empirically true of representative space, then

> there would be no geometry, there would be no space, because we should have no more interest in classifying the internal and external changes . . ., in distinguishing changes of state from changes of position. (Poincaré 1906a, p. 52)

Hence there can be no question of any relativism in Poincaré's conception of representative space, which he construed in a classically empiricist way: perception imposes—as it were, from without—the structure of this space on us. He seems however to contradict himself when he claims that experience singles out no specific mathematical geometry: no geometry is true to the exclusion of all others since all geometrical axioms are *conventions*, disguised definitions of a set of primitive predicates (Poincaré 1902, p. 50). All groups and all geometries pre-exist, potentially, in our minds; as for experience, it only shows us which of these structures it is most convenient for us to adopt (Poincaré 1902, p. 87; and 1906a, p. 52). With respect to theoretical geometry, Poincaré thus appears to be a throughgoing relativist. On closer examination, his theses however prove to be acceptable to any *broad-minded* realist. For example, let us consider Gauss's proposed experiment concerning the angles of a triangle ABC whose sides are formed by light-rays. Even though the angles ∠A, ∠B, and ∠C might not add up to π radians, Poincaré maintains that we should still have at least two solutions open to us: either we stick to the Euclidean Axioms and regard the sides AB, BC, CA as having been bent by some field of forcce, or else adopt a non-Euclidean geometry according to which AB, BC, and CA are geodesics, that is: shortest distances between given pairs of points (see below, 3.6). Using a more modern terminology, it can be claimed that we would be treating different geometries as formal calculi for each of which real space provides some model. These physical interpretations will of course vary from one calculus to the next; for example: the light-beam AB will not represent a geodesic in all of the envisaged (Riemannian) geometries (see below, 3.4). Poincaré however asks: what difference does it make whether AB is *called* a 'curved' line or a 'straight' one? After all, words are unimportant and only theories count because they convey structure: they express relations which physical objects bear to one another (see below, 3.6). This point was emphasized in *The Value of Science,* where Poincaré maintained that no matter what geometry we adopt, the structure of real space must be preserved.

> The straights of these men will not be our staights, but they will have among themselves the same relations as our straights to one another. It is in this sense I say their geometry will be ours. (Poincaré 1906a, p. 40)

Such passages, which will be re-examined later, are not easy to follow. Their interpretation is made even more difficult by Poincaré's claim that "Euclidean Geometry therefore has nothing to fear from fresh experiments" (Poincaré 1902, p. 73). This statement can be understood in at least three different ways. First, it may mean that the geometrical postulates, being testable only in conjunction with other assumptions, are empirically unfalsifiable, so that we can always cling to Euclidean Geometry by shifting the blame onto the associated physical hypothesis (Poincaré 1902, p. 73; also see below, 3.6). This is a logical point whose validity was later acknowledged by Einstein (1934a, pp. 122–23). Secondly, it is clear that Poincaré intended to go beyond mere logic: he maintained that no matter how complex our physics might become, Euclid's system would always prove more convenient than alternative geometries. This thesis was refuted by Einstein's General Theory of Relativity. Thirdly, Poincaré appeared also to be claiming that Euclidean Geometry is irrefutable because it defines the very methods of measurement used in physical experiments; these cannot—as it were—be turned against the system which made them possible in the first place. He wrote: "This convention defines both space and the perfect instrument" (Poincaré 1913a, p. 18). We have here an unmistakeably Kantian type of argument: objective experience presupposes and hence can neither confirm nor falsify Euclidean geometry. This is however a logically unsound claim actually undermined by any *reductio ad absurdum* proof: one can legitimately refute p by first assuming p and then inferring from it a logical or experimental falsehood.

It seems to me that Poincaré was led to these paradoxical conclusions by a conflation of mathematical modelling with physical semantics. He took geometry to be a deductive system *taken together* with a physical interpretation, which is *suggested* but not imposed by experience. His argument seems to run along the following lines: we can always choose the interpretation of a consistent system in such a way that all of its axioms are verified, so that no conflict can arise between any theorem on the one hand, and the 'results of experience' on the other. This is however true only of mathematical models. In the case of *physical* geometry, a number of objections can be raised. Even if they could be confronted with their intended referents—and later we shall see why they cannot—the axioms can never be strictly verified; this is so both because they consist of universal propositions and because they apply to experience only approximately; that is: within the limits prescribed by the precision of the available instruments. Thus two bodies may appear to have the same length when measured by one but not by another—more precise—method. Although our formal geometrical system remains fixed, its alleged physical model is therefore altered by the introduction of more accurate methods of measurement. The latter are expected to behave like the ideal instruments mentioned in the above quotation; but this is precisely the kind of expectation which further experimentation often frustrates. For example, the instruments developed towards the end of the nineteenth century enabled scientists to test the addition law of velocities applied to the earth frame and to light waves propagated in a motionless ether. This law had been 'verified' on innumerable occasions; which however did not prevent Michelson's results from undermining Newtonian kinematics (see below, Chapter 4).

Poincaré himself describes a situation in which, by his own account, Euclidean Geometry would prima facie be falsified. He imagines a universe in the form of a sphere, where the temperature decreases from some positive value at the centre towards absolute zero at the boundary. Thermal equilibrium is supposed to be instantaneously attained. In such a world, bodies do or do not obey Euclidean laws according to whether they are non-dilatable or highly-dilatable (see below, 3.4). Poincaré surprisingly concludes that in such a world, Euclidean Geometry would be both refuted and verified; which allegedly shows the absurdity of holding any geometry to be a testable hypothesis (see Poincaré 1902, p. 83). I need not insist on the logical oversight: in such a world, Euclidean geometry would be both falsified and corroborated, but not verified; for no amount of positive experimental evidence suffices to establish a synthetic universal proposition. The only way I can make sense of Poincaré's conclusion is by construing it as asserting that no physical phenomena can refute any of Riemann's whole spectrum of geometries. We shall later have to define the sense in which this thesis can be vindicated (see below 3.6).

Even more paradoxical is Poincaré's answer to the dimensionality question. We have seen how, in true empiricist fashion, he constructs our visual and tactile spaces which turn out to be three-dimensional and isomorphic to each other. Yet he surprisingly claims that the mind is free to ascribe any number of dimensions to *geometric* space. Since dimensionality is a discrete and finite quantity, he can no longer appeal to the approximate character of perceptual experience. He insists, moreover, that representative space has topological properties and that dimension is one of them. It therefore looks as though we finally have an invariant, namely 3-dimensionality, which, being directly given in experience, cannot be ignored. Poincaré rejects this conclusion through referring to Hertz's works, where the behaviour of a system of n particles is described as a generalised motion in a 3n-dimensional space:

> The simple fact that Hertz was able to conceive it as more convenient than our own habitual hypotheses suffices to prove that our ordinary ideas and, in particular, the

> three dimensions of space are in no wise imposed upon mechanics with an invincible force. (Poincaré 1906a, p. 69)

In this passage, Poincaré seems again to conflate the two questions of logical possibility and of empirical truth. He rightly maintains that we can 'think' in more than three dimensions, in other words that 3-dimensionality is neither an analytic nor a synthetic apriori property of space. This does not however mean that because of the choice of an abstract 3n-space, each separate point in Hertz's system has suddenly ceased to be determined by three coordinates.

Our next step is to try and reconcile these conflicting attitudes. A quotation from *Science and Method* sheds some light on this problem:

> Our distribution board, then, is only the translation of an assemblage of exterior facts: if it has three dimensions, it is because it has adapted itself to a world having certain properties; and the most important of these properties is that there exist natural solids, which are clearly displaced in accordance with the laws we call laws of motion of unvarying solids. (Poincaré 1908, p. 114)

Poincaré unequivocally expresses his belief in the existence of a mind-independent and structured external world, which will be denoted by *E*. We have no direct access to *E*, some of whose events are however reflected in the domain *R* of our sense-experience. Poincaré accepts Kant's thesis that the elements of *R*, being processed and organised by the mind, are not at all the raw sense-data which some empiricists take them to be. Finally, there exists a third domain *M*, consisting of purely abstract or mathematical entities. Geometry and representative space belong respectively to *M* and *R* which—taken together—form the intersubjective world of science. According to Poincaré, the choice of a dimension is a response of *M* to challenges posed by *E* and mirrored by *R*. Thus 3-dimensionality is a fixed feature of *R*, but with regard to theoretical geometry, *M* has at its disposal several mutually inconsistent options each of which is compatible with the results obtaining in *R*. One among these alternatives, namely the most convenient scientific hypothesis, will also prove to be the most effective solution in the sense of possessing the highest survival value (Poincaré 1958a, p. 114). Although no geometry adopted in M is empirically refutable, it can be undermined by events occurring in *E* and reflected in *R* (see below, 3.6). A conflict can thus occur between *R* and *M* on the one hand, and *E* on the other; where this conflict takes the form, not of a contradiction, but of the elimination of a species incapable of developing an effective solution to problems posed by its environment.

Let us now turn to Poincaré's philosophy of geometrical, as distinct from representative, space.

3.3 Poincaré's Half-Plane and Trivial Semantic Conventionalism (TSC)

In his two classics on space-time theories, *Philosophical Problems of Space and Time* and *Geometry and Chronometry in Philosophical Perspective*, Adolf Grünbaum examines a position which he calls "Trivial Semantic Conventionalism" (TSC) (Grünbaum 1973, p. 27). Grünbaum rightly rejects TSC, which can be described roughly as follows: taken by itself, no theory fixes the referents of its descriptive terms; these will consequently be given arbitrary meanings, provided the latter satisfy the theory's postulates; since the geometric predicates figure among the primitive terms of physical hypotheses, they will have automatically received such 'conventional' meanings. Geometric conventionalism is therefore subsumed by the meaning-conventionality of the descriptive vocabulary of any language.

TSC is trivial only in the sense of trivially following from the Skolem-Löwenheim theorem which, in itself, is far from trivial: it asserts that every syntactically consistent first-order theory expressible in a denumerable language has a countable model; if the theory furthermore has at least one infinite model, then for *every infinite cardinal* $\aleph_\alpha$ it will possess at least one model of cardinality $\aleph_\alpha$. Thus it is hardly surprising that non-Euclidean systems like synthetic hyperbolic geometry (SHG) should possess models in $\Re^3$, in $\Re^2$, in $\Re$, and even in ω (where $\Re$ denotes the real line and ω the set of natural numbers). Although Grünbaum attributes TSC to both Eddington and Putnam, the origins of this position go back to the founder of geometric conventionalism, Poincaré himself (Grünbaum 1973, pp. 24, 33–34). We should however remind ourselves that Poincaré's philosophical views, expressed at different times and under very diverse circumstances, were not always consistent, which is why a rational reconstruction of his philosophy becomes obligatory.

In the Chapter of *Science and Hypothesis* entitled 'Non-Euclidean Geometries', Poincaré constructs a so-called dictionary for translating SHG into the language of the 3-dimensional Euclidean continuum. For the sake of simplicity, I shall give the corresponding 'translation' of SHG into what

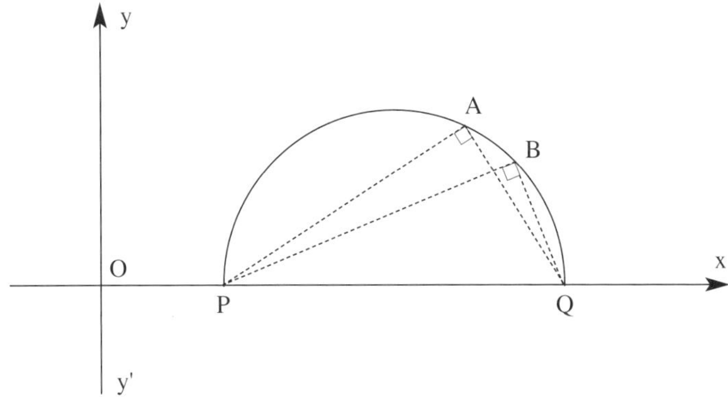

FIGURE 1

came to be known as the Poincaré Half-Plane, which will henceforth be denoted by Δ (see Figure 1).

H-Plane	Upper Half Plane Δ (more precisely : $\Delta = \{(x,y): x \in R,\ y>0\}$.
H-point	Point (more precisely: $(x,y),\ y>o$).
H-line	Vertical in Δ or semi-circle centred on some point of the X-axis (i.e. either $\{(a,y) : y>0\}$ or $\{(x,y): (x-a)^2 + y^2 = k^2,\ y>o\}$).
H-angle	Angle (i.e. $\cos^{-1}(d\underline{r} \cdot \delta\underline{r}/\lvert d\underline{r}\rvert \cdot \lvert\delta\underline{r}\rvert)$).
H-circle	Circle (i.e. $\{(x,y): (x-a)^2 + (y-b)^2 = k^2,\ 0<k<b\}$)
H-distance between 2 points	Logarithm of the anharmonic ratio (i.e. of the cross-ratio) of these 2 points and of the intersections of the X-axis with the circle passing through these 2 points and cutting the X-axis orthogonally (i.e. $\partial(A,B) = \lvert \mathrm{Log}(A,B;P,Q)\rvert = \lvert \mathrm{Log}(\underline{AP} \cdot \underline{BQ}/\underline{AQ} \cdot \underline{BP})\rvert$, where, for any 2 points U and V in the plane, $\underline{UV}$ is the complex number represented by the vector joining U to V).

In the above: H-plane, H-distance, and so on, stand respectively for: hyperbolic plane, hyperbolic distance, and so forth. Note that none of the analytic expressions written between brackets in the right-hand column figures in Poincaré's presentation. (They have been added for reasons which will be explained later).

Poincaré maintains that the above 'dictionary' enables us to translate all the axioms of SHG into theorems of ordinary Euclidean geometry. He goes on to claim that the difference between the two modes of speech, the hyperbolic mode and the Euclidean one, is on a par with the difference between two natural languages like German and French; also that choosing between two geometries is like choosing between two unit or two coordinate systems. This was later echoed by Schlick, who maintained that:

> we shall accept the result that space itself in no wise has a form of its own; it is neither Euclidean nor non-Euclidean in constitution, just as it is not a peculiarity of distance to be measured in kilometres and not in miles. (Schlick 1920, p. 3)

Despite Poincaré's limpid style and any feeling we might experience of having fully grasped his thesis, calling the above table a 'dictionary' is not only ambiguous but also gravely misleading. In order to be understood by a wide public, Poincaré laudably tried to avoid all technical jargon in his philosophical writings. This may nonetheless have given rise to misunderstandings on the part, not only of laymen, but also of philosophers and of scientists. Furthermore, the model Δ appears to be composed of a number of very disparate components: we do not understand why half-infinite *verticals and semi-circles normal to X'X*—but *no oblique* straight lines—should count as H-straight; why every circle lying *above* X'X should be regarded as H-circular, whereas no circle intersecting X'X can be H-straight or H-circular. Before confronting the definition of the H-distance, we also acquire the conviction that Poincaré is straightforwardly translating SHG into SEG, that is: into Synthetic Euclidean Geometry. But by the time we finish reading the expression of the H-distance, the game is up: the author unexpectedly invokes such sophisticated concepts as those of the logarithm and of the cross-ratio of complex numbers. We suddenly realize that we are dealing, not with a mere 'translation' of SHG into SEG, but with the construction of a model of SHG in the complex plane. (This is one reason for entering the appropriate analytic expressions into the right hand columns of p. 75 above.) Needless to say, we can turn the plane into a 2-dimensional Euclidean space by defining $ds_e^2 = dx^2 + dy^2$; but we need add no specifically 'Euclidean' axioms to that part of set theory which is normally presupposed by complex analysis.

We can now easily check that all the axioms of SHG 'translate' into theorems about complex numbers. It can moreover be shown that through any point outside any H-line, *l*, there exist infinitely many H-lines which are parallel to, that is: which do not intersect *l*. We have however mentioned that such proof-checking comes nowhere near any real understanding of Poincaré's model; for there is as yet no central idea binding the various components of Δ. We have no insight into why and how this model works but are left with the vague impression that Δ must be the transform of a more unified structure. Our only clue is the fact that verticals are treated on a par with certain semi-circles and that all H-straights are normal to X'X. This brings to mind specific conformal mappings, for instance inversions and Möbius transformations, which can interchange circles and straight lines. It will turn out that this is indeed the correct way of approaching Poincaré's construction of his various SHG-models; but we have so far had few clear indications pointing in this direction. (From an analytic viewpoint, Möbius transformations are preferable to inversions: Möbius Transformations—or MT's, for short—are holomorphic functions built by means of the usual algebraic operations; they furthermore preserve the magnitude *and* the sense of all angles. Thus, it comes as no surprise that Poincaré should have used MT's in preference to inversions. Note that the latter can anyway be viewed as components of the former).

We now turn to Poincaré's epistemological standpoint as expressed in *Science and Hypothesis*:

> Let us now take Lobatchewsky's theorems and translate them, by the aid of this dictionary, as we would translate a German text with the aid of a German-French dictionary. We shall then obtain the theorems of ordinary geometry. (Poincaré 1902, p. 42)

And:

> The geometrical axioms are therefore neither synthetic a priori intuitions nor experimental facts. They are conventions. Our choice among all possible conventions is guided by experimental facts; but it remains free and is only limited by the necessity of avoiding every contradiction . . . In other words, the axioms of geometry (I do not speak of those of arithmetic) are only definitions in disguise. What then are we to think of the question: is Euclidean geometry true? It has no meaning. We might as well ask if the metric system is true and if the old weights and measures are false; if Cartesian coordinates are true and polar coordinates false. One geometry cannot be more true than another; it can only be more convenient. Now, Euclidean geometry is and will remain the most convenient; . . . it is the simplest in

itself, just as a polynomial of the first degree is simpler than a polynomial of the second deree . . . (Poincaré 1902, p. 50)

This passage contains the thesis that differences between various geometries are on a par with those between different natural languages or between distinct coordinate systems, a view which is clearly indicated by Poincaré's use of the notion of a *translation* rather than of the *construction of a model.* This linguistic thesis is however highly suspect. Barring some nuances which are better expressed in one language rather than in another, two natural languages like German and French have roughly the same structure. A German-French dictionary is moreover supposed to preserve meanings irrespective of truth-considerations; whereas Poincaré's 'translation' alters the meanings of many familiar terms and is specifically designed to turn the postulates of SHG into theorems, into true sentences of analytic Euclidean geometry. In this way, we obtain an imbedding of SHG in the complex plane; but there is no question of SHG being equivalent to—or of constituting a substructure of—Euclidean geometry: the curvature of the latter still vanishes while that of hyperbolic geometry remains negative.

Translations between languages are therefore structure-neutral in a way in which a transition from one geometry to another is not. Yet the linguistic approach to alternative metrizations was adopted by many distinguished philosophers of science; for example by Reichenbach and—much more convincingly—by Adolf Grünbaum (see Reichenbach 1928, §§4, 8; also Grünbaum 1973, pp. 74–75). The interest and coherence of Grünbaum's position stem from his having clearly recognized that if various metrizations are to operate as alternative languages, then when considered prior to certain methodological *decisions*, geometrical notions such as those of congruence and curvature can possess no referents. Since Riemannian concepts are all expressible in terms of one metric tensor field, $g_{ij}(\underline{x})$, it follows that physical space must be metrically amorphous; a view which, as already mentioned, was also endorsed by Schlick: *we decree* the metric with which space is to be equipped; we then use this freely stipulated metric in order to express relations between *physical processes*,which alone have a claim to objective existence. Since different g_{ij}'s could have been chosen in order to describe the same facts, the difference between two metrics can legitimately be regarded as linguistic in character. Thus the core of Grünbaum's geochronometric conventionalism is the denial of the existence of any intrinsic metric.

Against this view, which is attributed both to Riemann and to Poincaré, I maintain that on a rational reconstruction of Poincaré's position, we find him holding geometry to be a theory about a mind-independent reality; more precisely: about a physical continuum which is not so much amorphous as intrinsically unobservable. Only through specifying the relations which space bears to concrete processes do we obtain a globally testable system. Thus only the *unobservability* of space enables us to choose arbitrarily any geometry we please. We shall however see that all these choices are not metaphysically on a par. Poincaré's thesis is therefore *epistemological* rather than ontological, which is evidenced by the following passage also taken from *Science and Hypothesis*:

> Experiments only teach us the relations of bodies to one another. They do not and cannot give us the relations of bodies and space, nor the mutual relations of different parts of space. (Poincaré 1902, p. 79)

Let us now examine Poincaré's claim that his 'translation' turns the axioms of SHG into theorems of Euclidean geometry. It looks as though a set of self-evidently false statements were changed, by a bizarre redefinition of their primitive concepts, into true sentences. This interpretation is a clear instance of TSC, which can thus be read into some of Poincaré's theses. TSC was consequently equated, not only by Eddington and by Putnam but also by Popper, with conventionalism pure and simple; and Popper can be excused for having attributed to conventionalists the following 'immunising' stratagem: when faced with an empirical refutation, a scientist can legitimately so redefine his terms that his postulates become automatically verified (Popper 1959, §16; also Albert 1975, §§5, 15). That this can always be done—in the case of any syntactically consistent system—is guaranteed by the Skolem-Löwenheim theorem; and this is one reason why, apart from regarding as scientific only falsifiable propositions, Popper insisted on our taking a firm methodological stand: that of never resorting to *conventionalist stratagems* when confronting any kind of criticism (Popper 1959, §20).

TSC will be rejected as a suitable rendering of Poincaré's position, but for reasons different from those of Grünbaum. I propose to show that TSC is not so much trivial as largely irrelevant: it fails to perform the function assigned to it both by its critics and by its proponents; namely that of allowing us to hold fast to certain beliefs, come what may. Consider a hypothesis H which may be a theory about physical space; and suppose that H has—in one way or another—been experimentally undermined. The only method by which TSC can help us to retain H is by altering the referents of some terms occurring in 'H'. But this clearly means that we are now envisaging a new *proposition* expressed by 'H', thereby admitting that the old H has been refuted or rejected. We may thus have succeeded in saving the string of symbols 'H' but certainly not the proposition previously denoted by 'H'. Hence, even if it could succeed, TSC would still have failed in that, in order to retain a proposition, it cannot but alter it.

Let me now show that TSC will not succeed—except at prohibitive cost—in saving even the bare *syntactic* expression, 'H'. We have, so far, considered exclusively theoretical terms; for note that geometrical notions like those of *congruence* and *curvature* are non-observational. (Should we accept the thesis of the amorphousness of space, then geometrical terms will be non-referring and hence trivially non-observational.) The realist naturally assumes that theoretical terms may denote, or at least reflect certain aspects of some transcendent reality; but he must also accept that we cannot leap out of ourselves and get hold of the referents of such terms in order to alter their relation to our language. The objects denoted by theoretical primitives are known not by acquaintance but by description, which is to say: through the medium of certain conjectures. Thus we can tamper with the wording of our theories but not with their semantic linkage to the world; for this linkage is in turn grasped by means of theory alone. Putting it slightly differently, though our hypotheses may refer to a mind-independent reality, it is only at the observational level that we have any direct control over the way in which laws latch on to that reality. As far as theoretical propositions are concerned, we have no choice but to handle them in purely syntactical fashion—to treat them as though they were empty formulas. (At first sight, this thesis may seem trivial, but when conjoined with the phenomenological view of observation—with the thesis that basic statements refer exclusively to our immanent mental states—then it has been regarded by some as too radical (see above, 1.7 and 1.8).)

At this point, it should be recalled that if a relation of the form H } p holds, then it must do so independently of the meanings of any descriptive terms occurring either in 'H' or in 'p'. After all, meaning-independence is the hallmark of deductive validity. Unless we modify our logic, then no matter what meanings we attach to our primitives, the consequences of H will remain *syntactically* unaltered. Suppose now that p is an experimentally falsified empirical prediction. As long as p is regarded as false, then unless we change our logic, we shall be unable to save H from refutation. Should we want to retain H, then only one solution remains: we must so tamper with the meanings of the terms occurring in 'p' that our experimental findings no longer undermine p. But since p is empirical, all of its terms should be strictly observational. We therefore have no choice but to alter the denotations of some *empirical* predicates; though certainly feasible, this move will automatically modify the *proposition expressed* by 'p' as distinct from the *syntactic entity* 'p'. We may thus run the risk of turning p into a tautological sentence; p will then cease to have not only any theoretical, but also any practical value. Arguing against Edouard Le Roy, Poincaré wrote:

> Recall the first example he [Le Roy] has given. When I say: phosphorous melts at 44°, I think I am enunciating a law; in reality, it is just the definition of phosphorous; if one should discover a body which, possessing otherwise all the properties of phosphorous, did not melt at 44°, we should give it another name, that is all, and the law would remain true. . . .
>
> It is clear that if laws were reduced to that, they could not serve in prediction; then they would be good for nothing, either as means of knowledge or as principles of action. (Poincaré 1906a, p. 123)

This is one reason why Poincaré insisted on observational predicates being given meanings fixed in advance of theory. In order to give a rational reconstruction of Poincaré's position, one should further require such meanings to be strictly phenomenological, that is: to be immanent in the sense of describing our private mental states (see above, Chapter 1). To these conclusions it might be objected that all geometric concepts are theoretical, so that modifying the referents of terms occurring in some empirical consequences of H in no way affects the meaning of any geometry, G say. Thus G can be retained, both form and meaning intact, even though the sense of some observational predicates might have been altered. But by the same token, it is not the semantic tinkering with geometric terms—as recommended by TSC—but something quite different which would have saved G. This 'something' must clearly have a syntactical character (see below, 3.7).

One last possibility—which could naturally not be envisaged by Poincaré himself—ought to be mentioned; namely that of so modifying our logic that p no longer follows from H. Such a move is tantamount to changing the meanings of certain connectives, which will in turn induce a change in the structure of our whole system of knowledge. Since the connectives are bound to occur in the formulation of the geometry, G, the meaning of G will have been modified. Some old theorems will moreover cease to be logically derivable from the geometrical axioms, even though the latter might have remained formally unaltered. It would therefore be highly misleading to claim that G, as a *proposition* or as a *class of propositions*, had been kept intact. Thus TSC in no way enables us to retain G as well as the prescribed empirical content of some scientific theory incorporating G.

3.4 Poincaré's Disk Model

We have seen that although we can check that it works as a model of SHG, Poincaré's half-plane Δ possesses an unintelligible structure. It consists of various components not bound together by any unifying heuristic idea. Poincaré's second model arises from quasi-physical considerations and is consequently more convincing than the half-plane:

> Suppose, for example, a world enclosed in a large sphere and subject to the following laws. The temperature is not uniform; it is greatest at the centre and gradually decreases as we move towards the circumference of the sphere, where it is absolute zero. The law of this temperature is as follows: if R be the radius of the sphere and r the distance of the point considered from the centre, the absolute temperature will be proportional to (R2–r2). Further, I shall suppose that in this world all bodies have the same coefficient of dilatation, so that the linear dilatation of any body is proportional to its absolute temperature. Finally, I shall assume that a body transported from one point to another of different temperature is instantaneously in thermal equilibrium with its new environment. (Poincaré 1902, p. 65)

For the sake of simplicity, I shall again confine my attention to the 2-dimensional version of this model, in other words to the unit disk in the plane. The historical evidence, anyway, suggests that Poincaré constructed his disk-model long before he thought about the half-plane which now bears his name. In 1881 he wrote:

> I call the circle with centre at the origin and radius 1 the fundamental circle; the hyperbolic group is the group of operations which change z to (az+b)/(cz+d) (where a,b,c,d are constants) and which leave the fundamental circle fixed; a discontinuous group is any group which does not contain an infinitesimal operation (an operation which changes z to a quantity infinitely close to z); a Fuchsian group is any discontinuous group contained in the hyperbolic group.
>
> I call any uniform function of z which is unchanged under the operations of a Fuchsian group a Fuchsian function. (Stillwell 1985, pp. 47–48)

Denote by *w* the circle $x^2 + y^2 = R^2$, where R is some positive number; and let Ω be the interior of *w*. Suppose that a temperature gradient instantaneously affects the dimensions of all bodies in exactly the same way. More precisely: assume that as we move towards the circumference *w*, all measuring rods are shortened by the factor $h(R^2–r^2)$, where h is some positive constant. Hence the measured distance ds between two neighbouring points will be equal to $ds_e / [h(R^2–r^2)]$, where ds_e is the ordinary Euclidean interval between the points in question. Thus:

$$ds_e = \sqrt{(dr^2 + r^2 \cdot d\mu^2)} = \sqrt{(dx^2 + dy^2)}.$$

Hence: $ds = \sqrt{(dr^2 + r^2 \cdot d\mu^2)} / [h(R^2 - r^2)] = \sqrt{(dx^2 + dy^2)} / [h(R^2 - (x^2 + y^2))]$, where r and μ are the polar coordinates of the point P(x,y). I.e. $x = r.\cos\mu$, $y = r.\sin\mu$.

In order to determine the geodesics of our (non-Euclidean) geometry, we have to minimize the integral:

$$\int ds = \int F(r,\mu').dr, \text{ where } \mu' = d\mu/dr, \text{ and } F(r,\mu') = \sqrt{(1 + r^2\mu'^2)} / [h(R^2 - r^2)].$$

By the Euler-Lagrange equations: $d(\partial F/\partial \mu')/dr - \partial F/\partial \mu = 0$. But since μ does not occur explicitly in F, we have: $\partial F/\partial \mu = 0$; hence $d(\partial F/\partial \mu')/dr = 0$. That is:

$$k = \partial F/\partial \mu' = r^2\mu' / [h(R^2 - r^2) \cdot \sqrt{(1 + r^2\mu'^2)}],$$

where k is some constant. Squaring, solving first for μ'^2 and then for μ', we obtain: $\mu' = hk(R^2 - r^2) / [r\sqrt{(r^2 - (hk)^2(R^2 - r^2)^2)}]$.

Putting $r = Ru$ and noting that $\mu' = d\mu/dr = (1/R).(d\mu/du)$, we infer:

$$\mu = \int h'.du.(1 - u^2) / [u \cdot \sqrt{(u^2 - h'^2(1 - u^2)^2)}] =$$
$$\int h'((1/u) - u) \cdot du / [u \cdot \sqrt{(1 - h'^2((1/u) - u)^2)}],$$

where we define: $h' = hkR$. This suggests the change of variable: $u = e^{-v}$. Thus:

$$\mu = - 2h'\int dv \cdot Sh(v) / [1 - 4h'^2Sh^2(v)]^{1/2} = - 2h'\int dv \cdot Sh(v) / \sqrt{[1 + 4h'^2 - 4h'^2Ch^2(v)]}.$$

Since $Sh(v) = d[Ch(v)] / dv$, we can integrate directly, thus obtaining:

$$\mu = - \sin^{-1}[2h'Ch(v) / \sqrt{(1 + 4h'^2)}] + \beta,$$

where β is some constant angle. Hence:

$$\sin(\beta - \mu) = 2h'Ch(v)/\sqrt{(1 + 4h'^2)} = h'(e^v + e^{-v}) / \sqrt{(1 + 4h'^2)} =$$
$$h'[(1/u) + u] / \sqrt{(1 + 4h'^2)} = h'[(R/r) + (r/R)] / \sqrt{(1 + 4h'^2)},$$

since, by definition: $r = Ru$.

Therefore $r \cdot \sin(\beta - \mu) = h'(R^2 + r^2) / [R\sqrt{(1 + 4h'^2)}]$.

Remembering that: $r^2 = x^2 + y^2$, $x = r.\cos\mu$, $y = r \cdot \sin\mu$ and $\sin(\beta - \mu) = \sin\beta.\cos\mu - \cos\beta.\sin\mu$, we finally obtain:

$(x - Rl \cdot \sin\beta/2h')^2 + (y + Rl \cdot \cos\beta/2h')^2 = (R/2h')^2$, where we put $l = \sqrt{(1 + 4h'^2)}$.

This equation represents the circle radius $(R/2h')$ and centre $(Rl.\sin\beta/2h', - Rl.\cos\beta/2h')$. Let us denote this circle by G.

We recall that *w* is the fundamental circle centre O and radius R. G is thus orthogonal to *w* since: $(Rl \cdot \sin\beta/2h')^2 + (Rl \cdot \cos\beta/2h')^2 = (Rl/2h')^2 = (R/2h')^2(1 + 4h'^2) = R^2 + (R/2h')^2$; that is: since the square of the distance

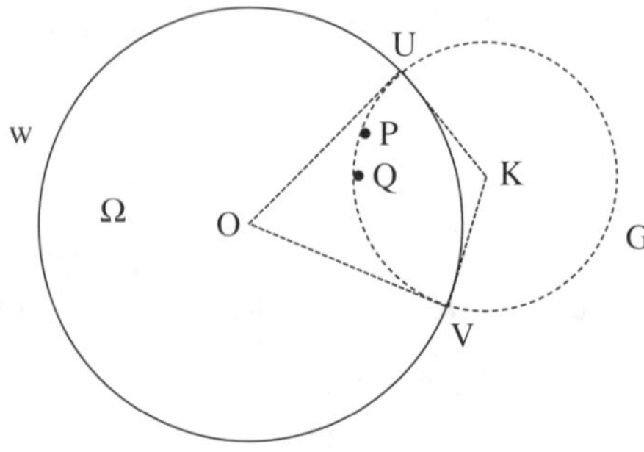

FIGURE 2

between the centres of G and of *w* equals the sum of the squares of the radii (see Figure 2).

Note that if k = 0, then μ' = 0; so we obtain the special case: μ = ß = constant, which represents a diamater of *w*. This is why it makes sense to call every circle and every straight line a *generalized circle*. Thus the geodesics of ds consist of all arcs of generalized circles which lie wholly within Ω and are orthogonal to *w* (see Figure 2).

In order to determine the distance between two points, P and Q, of Ω, it suffices to compute the integral $\int ds = \int\sqrt{(dr^2 + (r \cdot d\mu)^2)} / [h(R^2 - r^2)]$ along the geodesic connecting P with Q, that is: along the arc from P to Q which is normal to *w*. Denote by U and V the intersections of *w* with this arc. Though tedious, the above integral presents no serious difficulties; we obtain: $\partial(P,Q) = k' \cdot |L(P,Q;U,V)|$, where k' is the constant 1/(2Rh) and $\partial(P,Q)$ the distance $\int ds$ between P and Q.

Poincaré short-circuited this laborious method, first by using an MT in order to map Ω onto Δ, in other words onto the upper half of the complex plane; then by working exclusively within Δ. He proceeded as follows:

Let *M* be the class of all MT's, that is: of all complex functions of the form (az + b)/(cz + d), where ad – bc ≠ 0. It can be easily verified that *M* forms a group—that inverses and products of MT's are also MT's. MT's moreover possess the following basic properties:

[1] All MT's send generalized circles into generalized circles. An MT can thus transform a straight line into a circle and vice-versa.

[2] Every MT maps the interior of every generalized circle G (viz. the exterior of G) either wholly onto the interior or wholly onto the exterior of the transform of G.

[3] All MT's are conformal mappings: they preserve both the magnitude and the sense of all angles between curves.

[4] MT's preserve the anharmonic or cross ratio between any four points. In other words, for any MT f and for any complex numbers z_1,z_2,z_3,z_4, we have: $(z_1,z_2;z_3,z_4) = (f(z_1),f(z_2);f(z_3),f(z_4))$, where we define : $(x,y;\, z,u) =_{Def.} [(x - z) \cdot (y - u)] / [(x - u) \cdot (y - z)]$.

[5] We should finally note that the cross-ratio (x,y;z,u) is real iff x,y,z and u lie on some generalized circle.

In order to map Ω onto Δ, it therefore suffices to determine an MT *f* which both maps *w* onto the X-axis and sends at least one point z = x + iy of Ω into some point z' = x' + iy' of Δ; that is: we must have y' > 0. Put z' = *f*(z). In order to construct such a function *f*, we could for example require that: *f*(R) = R, *f*(–R) = – R and *f*(– Ri) = 0. Since *f* preserves all cross-ratios, it follows that: (z',R; – R,0) = (z,R; – R, – Ri). That is: (z' + R)R / 2Rz' = (z + R)R(1 + i)/2R(z + iR), which enables us to express z' in terms of z, and vice-versa. Thus:

(1) R(z + iR)/(iz + R) = z' = *f*(z); and:

(2) R(iz' + R)/(z' + iR) = z = f^{-1}(z') (see Figure 3).

Since *f*(iR/2) = 3iR and since 3iR lies above the X-axis, *f* maps Ω onto the upper half-plane Δ. Putting z = x + iy and z' = x' + iy', we infer from (2) above that:

$$x + iy = z = R(R - y' + ix') / [x' + i(R + y')] = [2R^2x' + iR(x'^2 + y'^2 - R^2)] / [x'^2 + (R + y')^2].$$

Equating real and imaginary parts:

(3) $x = 2R^2x' / [x'^2 + (R + y')^2];\ y = R(x'^2 + y'^2 - R^2)/[x'^2 + (R + y')^2]$.

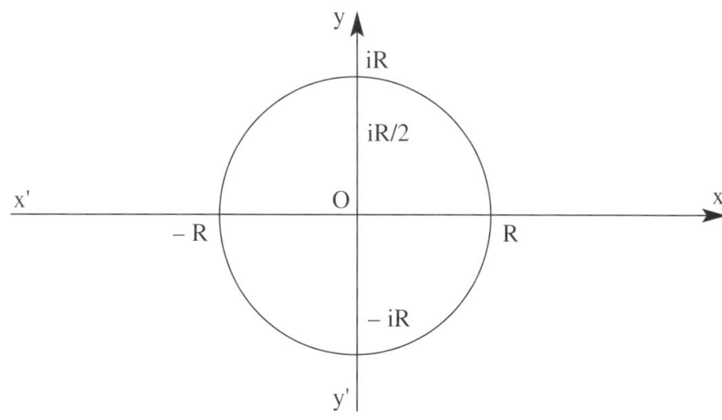

FIGURE 3

After a straightforward but tedious calculation, we find that:

(4) $k'\sqrt{(dx'^2 + dy'^2)}/y' = \sqrt{(dx^2 + dy^2)}/[h(R^2 - (x^2 + y^2))] = ds$ (by definition of ds), where, as above: $k' = 1/(2Rh)$.

Adopting the so-called passive view of the map *f*, we can look upon (x,y) and (x',y') as the coordinates, in two different systems, of one and the same point in H-space. Thus all the geodesics of Δ are of the form *f*[G], where G is a geodesic of Ω. We have moreover seen that G is the intersection with Ω of some generalized circle normal to *w*.

By (1)–(4) above, *f*[G] must be a generalized circle which both lies in Δ and is orthogonal to the transform *f*[*w*] of *w*. Remembering that *f*[*w*] = X'X, we conclude that *f*[*w*] is either a semi-circle centred on some point of X'X or else a vertical line lying wholly above X'X. Putting P' = *f*(P), Q' = *f*(Q), U' = *f*(U) and V' = *f*(V), then since all cross-ratios are invariant under *f*, we obtain:

(5) $\partial(P',Q') = \partial(P,Q) = k' \cdot |L(P,Q;U,V)| = k' \cdot |L(P',Q';U',V')|$

This result can also be achieved by direct calculation. Let $x' = b + a \cdot \cos\mu$, $y' = a \cdot \sin\mu$ be a parametric representation of the geodesic through P' and Q' (of the circle through P' and Q' which is centred on some point C of X'X). Thus: $dx' = -a \cdot \sin\mu \cdot d\mu$, $dy' = a \cdot \cos\mu \cdot d\mu$. So:

$$\partial(P',Q') = k' \cdot \int\sqrt{(dx'^2 + dy'^2)}/y' = k' \cdot \int a.d\mu/(a.\sin\mu) =$$
$$k' \cdot [L|\tan(\mu/2)|]^{Q'}_{P'} = k' \cdot L[|\tan(\beta/2)| / |\tan(\alpha/2)|] =$$
$$k' \cdot L[\underline{Q'V'} \cdot \underline{P'U'} / (\underline{Q'U'} \cdot \underline{P'V'})] = k' \cdot L(Q',P';V',U') = k' \cdot L(P',Q';U',V')$$

(see Figure 4).

The ease with which this integration can be carried out may well explain why Poincaré decided to concentrate on the domain Δ rather than on the

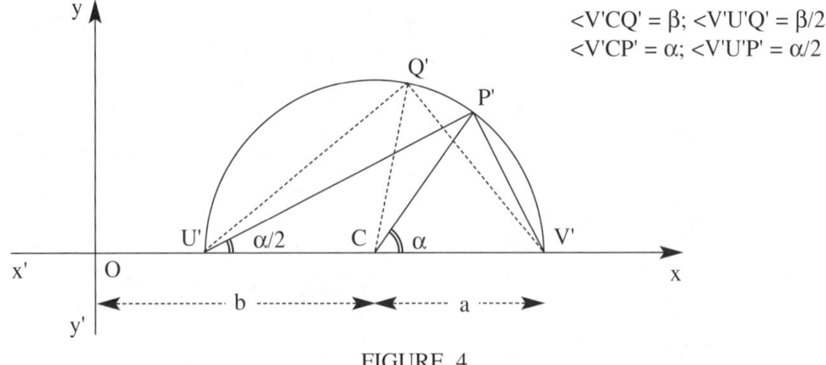

FIGURE 4

disk Ω. Note incidentally that the H-distance is obtained by multiplying ds_e by a single function of the coordinates, namely by $1 / [h(R^2 - (x^2 + y^2))]$ in the case of Ω, and by k'/y' in that of Δ. Euclidean angles will therefore be preserved in both hyperbolic models. Let us finally define an H-circle in Δ as the locus of all points P such that $\partial(Q,P) = k_o$, where Q is some fixed point and k_o a real positive constant. It easily follows that this locus coincides with an ordinary circle which lies above X'X, but whose Euclidean centre is distinct from Q.

Though equivalent to Δ, Poincaré's *circular world* Ω constitutes a vast improvement on the *half-plane* Δ. By envisaging a fictitious temperature gradient, Poincaré modifies the metric $ds_e = \sqrt{(dx^2 dy^2)}$ into $ds = \sqrt{(dx^2 + dy^2)}/[h(R^2 - (x^2 + y^2))]$. In strict Riemannian fashion, he then derives all the properties of Ω from this one expression of ds. The mapping *f* provides a rationale for all the apparently disconnected features of Δ, thus enabling us to understand how this second model works. But Poincaré's heuristic, as so far described, is not quite unitary. Two heterogeneous strands can be discerned within it: we are presented, on the one hand, with the usual Riemannian methods based on the expression of the distance ds and, on the other, with the group-theoretical approach advocated by Felix Klein in his *Erlanger Programm.* Nothing prepares the reader for the irruption, into the above deductive chain, either of the group *M* or of the holomorphic application *f* of Ω onto Δ. From Poincaré's own writings, we also know that he disapproved of Riemann's purely analytic viewpoint. He regarded it as too lax, since it licenses the construction of manifolds of variable curvature which do not deserve to be called *geometries of space.* According to Poincaré, an admissible space must have a constant curvature which guarantees the free mobility of all geometrical figures—these can, in such a space, be moved about without distortion.

> All depends, he [Riemann] says, on the manner in which the length of a curve is defined. Now, there is an infinite number of ways of defining this length, and each of them may be the starting point of a new geometry. That is perfectly true, but most of these definitions are incompatible with the movement of a variable figure such as we assume to be possible in Lie's theorem. These geometries of Riemann, so interesting on various grounds, can therefore never be anything but purely analytical, and would not lend themselves to proofs analogous to those of Euclid. (Poincaré 1902, pp. 47–48)

I therefore surmise that Poincaré based his approach to geometry on a transformation group from which he *subsequently derived* the infinitesimal expression of a metric. Should this conjecture turn out to be true, then it would show that Poincaré followed a single coherent heuristic throughout his geometrical researches. Only by digging back into history shall I however be able to vindicate this guess.

3.5 The Unity of Poincaré's Heuristic, or His Simultaneous Discovery of Fuchsian Functions and of Different Models for Hyperbolic Geometry

In the 1880s Poincaré developed his theory of Fuchsian Functions in order to integrate a whole family of differential equations. I shall not go into any details about the nature of these functions. Suffice it to say that there were good *nongeometrical* reasons for their introduction, and that they were defined through invariance properties under certain groups of MT's, that is: under specific subgroups of *M* (Stillwell 1985, pp. 47–50, 55–57).

We shall henceforth take R = h = 1 in equations (1)–(5) above. *w* will thus be the unit circle, Ω the unit disk, and *f* the function defined by:

(6) $f(z) = (z + i)/(iz + 1)$, so that: $f^{-1}(z) = (-z + i)/(iz - 1) = (iz + 1)/(z + i)$.

As above, Δ will denote the upper-half plane. Thus:

(7) $f[w] = X'X$ and $f[\Omega] = \Delta$, hence: $f^{-1}[X'X] = w$ and $f^{-1}[\Delta] = \Omega$.

Let us go back to the second quotation on p. 000, which is taken from one of Poincaré's earliest writings on Fuchsian functions. Following his own terminology, define

(8) *H* to be the set of all MT's which map each of *w* and Ω onto itself; that is: $g[w] = w$ and $g[\Omega] = \Omega$, for all $g \in H$.

Assuming that g and f belong to *H*, we have: $g[w] = w = f[w]$; hence: $fg[w] = f[w] = w$ and $g^{-1}[w] = g^{-1}g[w] = w$. Similarly: $fg[\Omega] = \Omega$ and $g^{-1}[\Omega] = \Omega$. Hence: $fg \in H$ and $g^{-1} \in H$, which establishes that

(9) *H* is a subgroup of *M*. This will be written as: $H \leq M$.

Poincaré refers to *H* as the hyperbolic group, and to a Fuchsian function as a complex mapping q defined in Ω and point-wise invariant under some (discontinuous) subgroup *F* of *H*; that is: q(z) = q(f(z) for all $f \in F$ and $z \in \Omega$; where $F \leq H \leq M$.

Although the structure of *F* need not concern us, that of *H* has great relevance to our geometrical considerations. Poincaré tells us how, in a sudden flash of intuition, he realized that *H* could be interpreted as the group of displacements of some non-Euclidean geometry. It should be remarked that such 'intuitions' are of the rational and not of the psychological or mystical kind. Felix Klein had after all established a strong link between geometry and group theory. It was also well-known that scalar functions are, by definition, invariant under all Euclidean displacements; for the latter can be viewed—in the passive mode—as changes of frames of reference. That *H* might constitute the group of direct isometries of some (non-Euclidean) geometry G was therefore not a far-fetched idea. But it took Poincaré's mathematical genius to determine G from the structure of *H*.

He used a method which he was later to apply to the construction of a Lorentz-covariant theory of gravitation. He took the simplest invariants of a given group—of *H* in the case of hyperbolic geometry and of the Lorentz group in that of gravitation—to be the building blocks of a hypothesis which was subjected to further adequacy requirements (see below, 4.8). The basic invariant features of MT's have already been listed on pp. 84–85. To these we have to add equations which follow from the definition of H: $g[w] = w$ and $g[\Omega] = \Omega$ for any $g \in H$. $g[\Omega] = \Omega$ tells us that Ω is the space on which the members of H operate; and $g[w] = w$ means that w should be regarded as a further invariant of *H*.

At this point, Poincaré simplified his task by using the function f in order to map the disk Ω onto the half-plane Δ (see (7) above). We have already noted that it is technically easier to deal with Δ than with Ω. Putting:

(10) $B = fHf^{-1}$,

let us show that every element of B maps each of X'X and Δ onto itself. Take $g \in B$; that is: $g = fhf^{-1}$ for some $h \in H$. Thus: $g[X'X] = f[h[f^{-1}[X'X]]] = f[h[w]]$ (by(7)) $= f[w]$ (by(8)) $=$ X'X (by(7)). We similarly establish that $g[\Delta] = \Delta$. Therefore:

(11) $g[X'X] = X'X$ and $g[\Delta] = \Delta$, for all $g \in B$.

In other words: B consists of all MT's, g, which both map Δ onto itself and send real numbers into reals. Such MT's can be singled out as follows. Let p, q, and r be any three distinct real numbers. Put: $g(p) = p'$, $g(q) = q'$, $g(r) = r'$, and $g(z) = z'$, where z is an arbitrary complex number. Thus p', q', and r' are real.

Being an MT, g preserves all cross-ratios; hence: $(z',p';q',r') = (z,p;q,r)$, that is: $(z' - q')(p' - r') / [(z' - r')(p' - q')] = (z - q)(p - r) / [(z - r)(p - q)]$. Solving for z' in terms of z, we obtain a relation of the form:

(12) $(az + b)/(cz + k) = z' = g(z)$, where a, b, c, and k are all real.

We must now make sure that g maps Δ onto itself. By Property [2], p. 000, it suffices to require that g(i) should belong to Δ. By (12): $g(i) = [(ac + bk) + (ak–bc)i]/(c^2 + k^2)$. We must therefore impose the further condition: $(ak – bc)>0$. Thus:

(13) $B = \{ (az + b)/(cz + k)$: a, b, c, k are real, and $(ak – bc)>0 \}$.

We can now determine the structure of *H* from (10) and (12) as follows. By (10):

(14) $H = f^{-1}Bf$.

Let $h \in H$; thus $h = f^{-1}gf$ for some $g \in B$. g will therefore be expressible in the form (12) given above.

Put:

(15) $h(z) = (uz+s)/(vz+t)$, where u, s, v, and t are complex constants.

Since $h = f^{-1}gf$, it follows from (6) and (12) that:

(16) $$\begin{bmatrix} u & s \\ v & t \end{bmatrix} = \begin{bmatrix} -1 & i \\ i & -1 \end{bmatrix} \bullet \begin{bmatrix} a & b \\ c & k \end{bmatrix} \bullet \begin{bmatrix} 1 & i \\ i & 1 \end{bmatrix}$$

By multiplying out and then equating corresponding entries in the matrices thus obtained, we infer: $u = -(a + k) + i(c - b)$; $s = -(b + c) - i(a - k)$; $v = -(b + c) + i(a - k)$; $t = -(a + k) - i(c - b)$. Hence:

(17) $s = v^*$, $t = u^*$ where, for any complex z, z* will henceforth denote the conjugate of z.

Through taking the determinants of both sides of (16), we obtain:

$ut - vs = 4(ak - bc)$. By (12) and (17), it follows that:

(18) $uu^* - vv^* = 4(ak - bc) > 0$, that is: $|u|^2 - |v|^2>0$ or $|u|>|v|$.

Therefore:

(19) $H = \{ (uz + v^*)/(vz + u^*) : |u|>|v| \}$.

As already mentioned, Poincaré needed to determine the Riemannian metric which fixes the geodesics of his geometry G. He actually set out to find some non-negative real-valued function ß(P,Q) such that ß is additive along every geodesic in Δ. Hence the geodesics of G in Δ must first be singled out. Note that if γ is to qualify as a geodesic, then for every $g \in B$, g[γ] should also prove to be a geodesic, for g[γ] can be regarded as a displaced image of γ. All these results had moreover to be derived from the structure of *B*. Once again, everything becomes more intelligible—if pragmatically less convenient—as soon as we consider Ω instead of Δ. For reasons of symmetry, the diameters of w must figure among the geodesics of Ω. (We know that one—and only one—geodesic is supposed to pass through two distinct points. So let ρ be a diameter of *w*, P and A two distinct points of ρ, and σ the geodesic through P and A. Were σ to be different from ρ, then, by symmetry, the reflection σ' of σ in ρ would be a geodesic distinct from σ and going through P and A, which is impossible. Hence $\rho = \sigma$.) But the elements of *H*, which are displacements of G in Ω, can transform the diameters of *w*, that is, segments of straight lines, into arcs of circles. Since all diameters are normal to *w* and since *w* is its own transform, these arcs must be orthogonal to *w* while remaining within Ω. In this way, we obtain, as class of the geodesics of Ω, the family Σ_o of all arcs of circles orthogonal to *w*; where the word 'circle' will henceforth denote any *generalized circle*, that is: any ordinary circle *or* ordinary straight line. The function *f* then transforms Σ_o into the set Σ of all arcs normal to X'X; that is: Σ consists of all the verticals of Δ together with the semi-circles centred on points of X'X.

We must now construct a real-valued nonnegative function ß such that:

(20) $\beta(P,Q) = \beta(g(P),g(Q))$, for all $g \in B$, $P \in \Delta$, $Q \in \Delta$;

and:

(21) $\beta(P,Q) = \beta(P,C) + \beta(C,Q)$ for any 3 points P, C, and Q of any member γ of Σ such that C lies between P and Q.

The fact that cross-ratios are invariant under all MT's ensures that condition (20) will be automatically satisfied once we succeed in expressing $\beta(P,Q)$ in terms of some cross-ratio. The latter is moreover known to be real for any 4 points lying on any circle, hence on any element of Σ (see [5] p. 85).

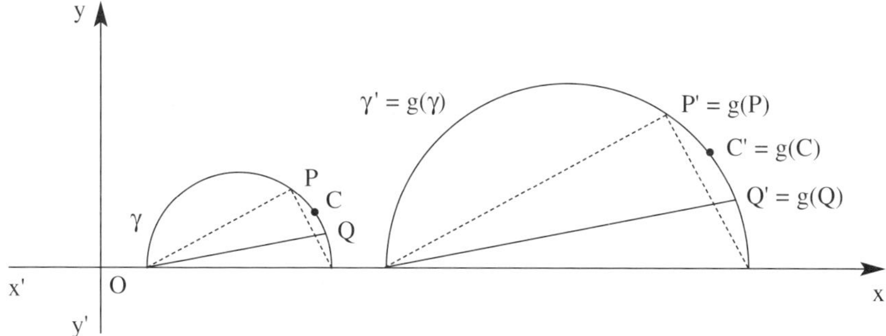

FIGURE 5

Consider the equation (P,Q;U,V) = PU.QV /(PV.QU), which holds by definition of the cross-ratio. Since P∈Δ and Q∈Δ, P and Q will be on the same side of X'X ; that is: of a diameter of γ (see Figure 5). Hence: (P,Q;U,V) ≥ 0; for the angles ∠UPV and ∠UQV have not only the same magnitude but also the same sense. Note that for any g∈B, g[γ] will be normal to g[X'X] = X'X. So let U' and V' be the intersections of g[γ] with X'X. Under g, the points U' and V' must correspond to U and V respectively; thus U',P',Q',C', and V' will lie on the transform γ' = g[γ] of γ. In other words: the semi-circle γ, which goes through P, Q, C and is normal to X'X, must be transformed by g into γ' = g[γ] on which the points: P' = g(P), Q' = g(Q) and C' = g(C) lie. Since g preserves angles, g[γ] = γ' will cut g[X'X] = X'X orthogonally, say in two points U' and V' which, being the intersections of γ' = g[γ] with X'X = g[X'X], must be the transforms of the intersections U and V of γ with X'X. That is: U' = g(U) and V' = g(V); hence: (P,Q;U,V) = (g(P),g(Q);g(U),g(V)) = (P',Q';U',V').

Note that P and Q uniquely determine γ, that is: the circle which passes through P and Q and is centred on some point of X'X. Thus if we define:

[P,Q] = (P,Q;U,V), then by the invariance of the cross-ratio:

[P,Q] = (P,Q;U,V) = (g(P),g(Q);g(U),g(V)) = (g(P),g(Q);U',V') = [g(P),g(Q)],

by definition of [g(P),g(Q)]. That is: [P,Q] is invariant under all g∈B (see Figure 5).

[P,Q] is however not additive but multiplicative along γ; for by the definitions of [P,Q] and of the cross-ratio:

[P,C].[C,Q] = (PU.CV/(PV.CU)).(CU.QV/(CV.QU)) = PU.QV/(PV.QU) = (P,Q;U,V) = [P,Q].

This result suggests that we should take the logarithms of all members of the above equations, thus obtaining:

(22) L([P,C]) + L([C,Q]) = L([P,Q]).

L([P,Q]) clearly satisfies all the conditions which, given the structure of the group B, must be imposed on the concept of distance. We can therefore define:

(23) β(P,Q) = |L[P,Q]| = |L(P,Q;U,V)|.

(The modulus signs are added in order to ensure that all the quantities occurring in the above equations remain non-negative).

Poincaré finally computed the distance between two neighbouring points, $z = x + iy$ and $z + dz = (x + dx) + i(y + dy)$. Through laboriously calculating $|L[z,z + dz]|$ and thereby neglecting second-order terms, he found that: $|L[z,z + dz]| = \sqrt{(dx^2 + dy^2)}/y$. To within a constant factor, this is the infinitesimal distance ds considered above (see Stillwell 1985, pp. 61–62).

Summing up: we can say that in pure geometry Poincaré used—right from the start—one and the same heuristic based on Klein's group-theoretical methods. He defined his Fuchsian Functions through their invariance properties under certain subgroups *F* of *H*; where *H* is a subgroup of the set *M* of all MT's (Möbius Transformations). "In a flash," he realized that *H* could be viewed as the family of displacements of some geometry G. G turned out to be hyperbolic, which is why *H* was called the hyperbolic group. The structure of *H* enabled Poincaré to find the Riemannian distance, which in turn determines the geometry G. In mathematical geometry, the algebraic approach was therefore central while Riemann's analytic methods remained, if not derivative, then at any rate secondary. In the physical sciences, this situation was naturally reversed. This is one reason why, in Poincaré's 1902, the hyperbolic distance was *anachronistically* presented as due to the existence of a temperature gradient which uniformly 'distorts' all measuring instruments. But more of this later.

3.6 Discovery or Invention of Universal Fields?

Let us go back to Poincaré's disk model as presented in Poincaré 1902. Despite his preference for group-theoretical methods, Poincaré must have realized that Riemann's approach to *physical* geometry, based as it was on the measurement of short distances, was bound to prove superior to its Kleinian rival. As already mentioned, this might explain why he described his model in terms of temperature gradients, of shrinkages and of thermal equilibria. We are told that if the inhabitants of the disk Ω use material rods in order to probe their spatial environment, they will be naturally led to adopt a hyperbolic metric as the foundation of their global geometry (Poincaré 1902, pp. 65–66). A creative mathematical physicist could however reinstate Euclidean geometry by introducing the temperature gradient described by Poincaré. This gradient typifies what Reichenbach later called a 'universal field', a force which cannot be shielded against and which affects *all* bodies in exactly the same way (Reichenbach 1928, §11).

The question arises as to whether such 'fields' were intended to be taken in a literal or in an exclusively metaphorical sense. As mentioned above,

Grünbaum equates the Riemann-Poincaré thesis with the view that physical space is metrically amorphous; which explains why we can dictate the metric and hence also the geometry of our choice. By the same token, we are debarred from asking what *causal* factors prevented some method of measurement from yielding our *arbitrarily* chosen metric ds. According to Grünbaum, all talk of 'universal fields' is but a metaphorical way of referring to the difference between the stipulated values of the $g_{ij}(\underline{x})$'s occurring in $ds^2 = g_{ij}(\underline{x}) \cdot dx^i \cdot dx^j$ and the $g'_{ij}(\underline{x})$'s which are empirically determined by the customary means. Only after *decreeing* the $g_{ij}(\underline{x})$'s and actually measuring the $g'_{ij}(\underline{x})$'s can we *define*: $F_{ij} = g'_{ij} - g_{ij}$, where F_{ij} represents a (fictitious) universal field. (Grünbaum 1973, Chapter 3A). For example: on Poincaré's disk, we can rely on our measuring rods to inform us that our 'natural' geometry is hyperbolic; it would however be absurd to wonder what physical forces so deformed our rods that our geometry ceased to be Euclidean. Such a question has a meaning only if, prior to any stipulation, space possessed an intrinsic metric—which is tacitly assumed to be Euclidean; but if space is in itself amorphous, then the differences between different metrics and hence between various geometries are linguistic in character.

> Indeed, after first commenting the factual basis of the existence of the Newtonian time congruence, we shall see that there are weighty pragmatic reasons for preferring that metrization of the time continuum. And these reasons will turn out to be entirely consonant with our twin contention that alternative metrizability allows linguistically different, equivalent descriptions and that geochronometric conventionalism is not a subthesis of trivial semantic conventionalism. (Grünbaum 1973, pp. 74–75)

Grünbaum not only powerfully argued his case but he also enlisted some of Poincaré's own writings in support of his linguistic thesis. As already conceded, Poincaré seemed at times to regard the difference between two geometries as one between two natural languages or two coordinate systems, a view which was later underwritten by Schlick (1920, Chapter 5). But it has also been remarked that over the years, Poincaré expressed inconsistent views about the role of conventions in physics. Hence the possibility of vindicating the following claim: on a rational reconstruction which is consistent not only with many of his writings but also with the spirit of his philosophy, Poincaré turns out to be a diehard structural realist; *diehard* in the sense of holding physical space to be equipped with an *intrinsic* though not directly calculable metric. More precisely: he took space to have a structural property mirrored by *the* geometry which is most coherently integrated into an empirically successful and maximally convenient scientific system. That he also mistakenly took this geometry to be Euclidean is irrelevant to his global philosophical position.

Let me start by pointing to a counter-intuitive consequence of the linguistic thesis, which cannot but conflate two vastly different problem-situations.

Consider any N-dimensional space, S. S is thus a point-set possessing a topology over which a family of invariants are defined as continuous real-valued functions. (The topology may have been generated by the invariants themselves, in the sense of being the weakest topology ensuring that all the latter are continuous maps.) Suppose that two metrics, or two second-order covariant tensor fields are defined on S. Take $\underline{x} = (x^1,\ldots,x^N)$ to be any suitable coordinatisation of S (in some region) and let $g_{ij}(\underline{x})$, $g'_{ij}(\underline{x})$ be the components of the two metrics in this coordinate system. Denote by $[g_{ij}(\underline{x})]$ and $[g'_{ij}(\underline{x})]$ the geometries determined by the metrics g_{ij} and g'_{ij} respectively. Then at most one of the following incompatible—though not exhaustive—situations can arise:

(a) There exists a well-behaved one-one mapping $(x^1,\ldots,x^N) = \underline{x} \rightarrow \underline{x}' = (x'^1,\ldots, x'^N)$ of S onto itself such that: $g'_{ij}(\underline{x}') = (\partial x^p/\partial x'^i) \cdot (\partial x^q/\partial x'^j) \cdot g_{pq}(\underline{x})$, where we make use of Einstein's summation convention. It follows that:

$$(ds(\underline{x}))^2 = g_{ij}(\underline{x}) \cdot dx^i \cdot dx^j = g_{ij}(\underline{x}) \cdot (\partial x^i/\partial x'^p) \cdot (\partial x^j/\partial x'^q).dx'^p.dx'^q = g'_{pq}(\underline{x}') \cdot dx'^p \cdot dx'^q = (ds'(\underline{x}'))^2.$$

We can adopt the so-called passive view of the map: $\underline{x} \rightarrow \underline{x}'$ and look upon $\underline{x}$ and $\underline{x}'$, not as two distinct points of S, but as two sequences of coordinates of the same point, P, in two distinct frames. Let us recall that, by definition, the geometrical properties are precisely those which do not depend on any specific coordinatisation. Hence: $[g_{ij}(\underline{x})] = [g'_{ij}(\underline{x}')]$. Since $\underline{x}$ and $\underline{x}'$ are mere labels, it also follows that: $[g_{ij}(\underline{x})] = [g'_{ij}(\underline{x})]$. With regard to the geometrical, as distinct from the metric structure of S, it can therefore be maintained that the choice between g_{ij} and g'_{ij} is on a par with that between two languages or two coordinate systems.

(b) Let $B^i_{jkl}(\underline{x})$ and $B'^i_{jkl}(\underline{x})$ be the curvature tensors built from the metrics $g_{ij}(\underline{x})$ and $g'_{ij}(\underline{x})$ respectively. Suppose that: $B^i_{jkl} \neq 0 = B'^i_{jkl}$. Since the vanishing of the curvature tensor expresses a *geometrical* property, it follows that: $[g_{ij}(\underline{x})] \neq [g'_{ij}(\underline{x})]$; for $[g'_{ij}(\underline{x})]$, but not $[g_{ij}(\underline{x})]$, describes a flat manifold.

Let us more specifically assume that N = 2 and that $[g_{ij}(\underline{x})]$ is hyperbolic while $[g'_{ij}(\underline{x})]$ is strictly Euclidean. Thus there exists a coordinate transformation: $\underline{x} \rightarrow \underline{x}' = (x'^1,x'^2)$, such that $\delta_{ij} = (\partial x^p/\partial x'^i) \cdot (\partial x^q/\partial x'^j) \cdot g'_{pq}(\underline{x})$, where δ_{ij} is Kronecker's delta. In other words: (x'^1,x'^2) are the ordinary Cartesian coordinates in the plane. Another transformation, $\underline{x} \rightarrow \underline{x}'' = (x''^1, x''^2)$ will yield: $g''_{ij}(\underline{x}'') = (\partial x^p/\partial x''^i) \cdot (\partial x^q/\partial x''^j) \cdot g_{pq}(\underline{x})$, where: $g''_{12}(\underline{x}'') = 0 = g''_{21}(\underline{x}'')$ and $g''_{11}(\underline{x}'') = g''_{22}(\underline{x}'') = 1/(x''^2)^2$. In this way, we obtain Poincaré's half-plane, Δ.

Using Poincaré's dictionary as presented on pp. 75–76, we can now translate $[g''_{ij}(\underline{x}'')] = [g_{ij}(\underline{x})]$ into $[\delta_{ij}] = [g'_{ij}(\underline{x})]$. But as already explained, the word 'translation' is here nothing short of a misnomer; for it denotes the construction of a model of $[g_{ij}(\underline{x})]$ within $\Re^2$. By definition, all the axioms of $[g_{ij}(\underline{x})]$ are thus sent into theorems of $[g'_{ij}(\underline{x})]$; but there can be no question of $[g_{ij}(\underline{x})]$ and $[g'_{ij}(\underline{x})]$ possessing the same geometrical structure: $[g'_{ij}(\underline{x})]$ is still flat while $[g_{ij}(\underline{x})]$ has a constant negative curvature.

In less technical terms, Grünbaum made a point similar to (a) above. He considered a table top being metrized in two different ways, each of which nevertheless yields the same (Euclidean) geometry. Thus he conclusively proved that Reichenbach and Carnap had confused the notion of a 'metric' with that of a 'geometry'; for whereas the former determines the latter, the converse does not hold. Grünbaum went on to claim:

> Assume now that the latter specification of the geometry G is not coupled with any information regarding F [= Universal Field]. Is it then correct to say that since this metrization provides no information at all about the coincidence behaviour of a rod under transport on the surface, it conveys no factual information whatever about the surface or about physical reality? That such an inference is mistaken can be seen from the following: depending upon whether the Gaussian curvature K associated with the stipulated gij is positive (spherical geometry), zero (Euclidean geometry), or negative (hyperbolic geometry), it is an objective fact about the surface that through a point outside a given geodesic of the chosen metric, there will be 0, 1, or infinitely many other such geodesics which will not intersect the given geodesic. (Grünbaum 1973, p. 97)

I have to admit that the last point made by Grünbaum, which is similar to (b) above, came as a complete surprise; for it runs counter to his thesis that conventionalism is made possible by the amorphousness of physical space; which is precisely why we can impose on space the metric and hence the geometry of our choice. But why then turn round and assert that various relations between *geodesics* constitute objective facts about the world? Once we postulate the absence of any intrinsic spatial metric, we are committed to regarding relations between *geometrical* entities as conventional, and the differences between various geometries as linguistic in character, that is: as differences between modes of description of the same facts. Putting it simplistically: when we adopt hyperbolic rather than Euclidean geometry, it is not space as such but our mode of speech which becomes curved; for space had no well-defined structure and geodesics did not exist prior to our methodological decision. It seems to me that at least on this point, Grünbaum ought to have agreed with Reichenbach. Yet he is—rightly, but so far unjustifiably—reluctant to accept Reichenbach's thesis concerning equivalent descriptions of identical states-of-affairs.

It should be noted that Poincaré's conception of the conventionality of geometry differs from those of Carnap and of Reichenbach. Poincaré subscribed to an epistemological notion of truth according to which only those statements possess truth-values which are at least partially decidable: such statements must be ascertainable at least as true or at least as false (see above, 1.4). All conventions and all geometrical hypotheses are accordingly devoid of truth-value; for since space is in principle unobservable, such hypotheses are neither empirically refutable nor verifiable, so they can be neither true nor false but more or less convenient. A convention's degree of convenience nonetheless functions as an index of verisimilitude and has therefore ontological import: the more convenient a system, the more faithfully it reflects the true nature of things. This is the crucial difference between logical empiricism on the one hand, and Duhem's and Poincaré's philosophies, on the other; for according to the latter, the unity and simplicity of scientific hypotheses bears a reference not only to the human mind but also to some transcendent reality (Duhem 1914, Chapter 2, §§4, 5).

Poincaré's views about the connection between geometry and physics can be rationally reconstructed as follows. All geometries specify relations between well circumscribed objects like points, geodesics, surfaces, distances and angles. Scientific theories then fix the relationship between these ideal entities on the one hand, and certain concrete physical processes, on the other. Physicists furthermore subscribe to a principle which can be called the *Generalized Principle of Inertia*, or GPI for short: other things being equal, some natural processes like the motions of free particles and the propagation of light take place along geodesics, that is: along some privileged configurations of the underlying geometry; hence only deviations from these privileged paths call for a causal explanation in terms, for example, of external forces. Consider two material objects or physical processes, which will henceforth be denoted by B_1 and B_2. For j = 1,2, let B^0_j be the region of space which,according to GPI, B_j should ideally occupy; that is: B_j would coincide with B^0_j were it not for certain distorting forces which may be either universal or differential or a mixture of both; where: differential $=_{\text{Def.}}$ dependent on the physical composition of B_j, universal $=_{\text{Def.}}$ nondifferential. Thus we have the following scheme:

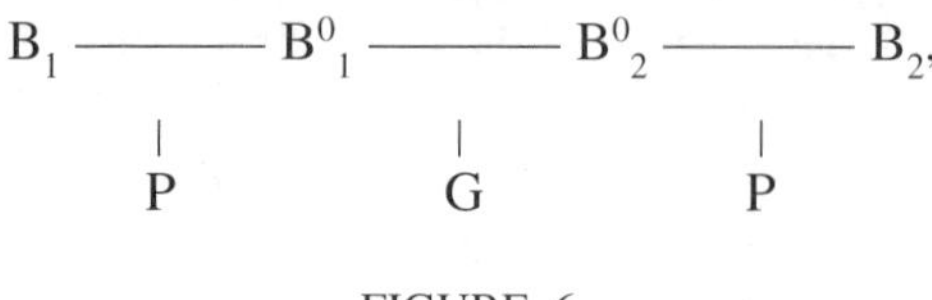

FIGURE 6

—where P denotes a physical theory and G the underlying geometry. Thus $G \wedge P$ yields a relationship between the two concrete physical entities, B_1 and B_2. Unlike G alone, the conjunction $G \wedge P$ is in principle falsifiable. In this way, we appear to be dealing with a typical Duhem-Quine situation. We are thereby led—or misled—into putting the system under test in the form $G \wedge P$; then into holding that were $G \wedge P$ to be refuted, we could in principle incriminate G or P or both. It therefore looks as though G and P were on a par. This is however far from being the case; for whereas G describes the properties of space-time and can consequently be made independent of P, P *presupposes* G both as a language and as a background theory. Thus the structure of a scientific system is adequately rendered, not by $G \wedge P$, but by $G \wedge P_G$, where the suffix indicates the dependence of physics on some prior geometry. We should add that the priority in question is of the logical and not of the genetic or heuristic kind.

Einstein appeared at one time to have undermined the primacy of geometry, whose dependence on energy seemed written into his celebrated field equations: $R^{ab} - (1/2)g^{ab}R = - kT^{ab}$. The left-hand side represents the geometry, and the right-hand side the physical content of space-time. It therefore looked as though geometry and physics were indissolubly mixed.

Let us however recall that the above equations were first proposed in the form: $R_{ab} = - k(T_{ab} - (1/2)g_{ab}T)$, where the second member involves both the metric and the energy tensor T_{ab}. We should moreover note that the basic condition imposed on the contravariant tensor T^{ab} is the vanishing of its divergence $T^{ab}{}_{:b}$, where the covariant derivative $T^{ab}{}_{:j}$ of T^{ab} depends on the affine connections and hence on the underlying geometry. The latter must therefore be *presupposed*. It follows that Einstein's approach did not jeopardize the primacy of geometry over other components of a physical system; for his equations could be viewed as merely providing a *method for calculating the metric*: after using Special Relativity in order to determine T_{ab}, one goes on to solve $R_{ab} = - k(T_{ab} - (1/2)g_{ab}T)$ for the unknowns g_{ab} in terms of the tensor (T_{ij}). Any solution will ensure that $T^{ab}{}_{:b} = (- 1/k) \cdot (R^{ab} - (1/2)g^{ab}R)_{:b} = 0$, for the second member vanishes identically.

It is worth noting that Einstein himself spoke of the left-hand side of $R_{ab} - (1/2)g_{ab} R = - kT_{ab}$ as being made of marble and of the right-hand side as consisting of low-grade wood. He also conceded that General Relativity (GTR) can be given a Euclidean formulation. In 'Geometry and Experience', he wrote:

> Thus the original, immediate relation between geometry and physical reality appears destroyed, and we feel impelled toward the following more general view, which characterizes Poincaré's standpoint. Geometry (G) predicates nothing about the behaviour of real things, but only geometry together with the totality (P) of physical laws can do so. Using symbols, we may say that only the sum (G) + (P) is subject to experimental verification. Thus (G) may be chosen arbitrarily, and also parts of (P); all these laws are conventions. All that is necessary to avoid contradictions is to choose the remainder of (P) so that (G) and the whole of (P) are together in accord with experience. Envisaged in this way, axiomatic geometry and the part of natural law which has been given a conventional status appear as epistemologically equivalent. *Sub specie aeterni* Poincaré is, in my opinion, right. (Einstein 1994, p. 258)

It follows from this quotation that Einsteinian gravitation can—in principle—be given a 'flat' formulation. As long as the signature of the metric is regarded as a distinctive feature of the (real) geometry, such flat space-time will naturally be of the Minkowskian rather than of the strictly Euclidean kind. All the same: Einstein had considerably shifted his position since the time, in 1911–12, when he held a special-relativistic theory of gravitation to be impossible. This change of heart may have been due to the fact that by 1921, Poincaré's work on the dynamics of the electron and his construction of a Lorentz-covariant theory of gravitation had become better known. Be it as it may, Einstein conceded that within the limits prescribed (say) by the signature of the metric, any geometry could in principle be decreed a priori. This may have been one reason why *physicists* like Eddington and Schrödinger went as far as claiming that only geometrical entities matter: asserting the presence of energy at some point $P(\underline{x})$ of space is another way of saying that $R_{ab} - (1/2)g_{ab}R \neq 0$. That is, geometry alone determines what parts of space-time have a physical content and what parts are empty of all but gravitational energy (Eddington 1923, p. 120; Schrödinger 1950, p. 99). This—admittedly extremist—attitude serves to emphasize the *logical* precedence of geometry over all other components of a scientific system. Logically speaking, we first settle on some definition of congruence by choosing the values of the $g_{ij}(\underline{x})$'s; these fix our geometry and enable us to express all laws within the framework provided by the metric, that is: by the freely chosen concept of congruence. This initial decision in no way debars us from subsequently so redefining our metric that our system becomes globally simpler or more unified. This is in fact what Poincaré proposed that we should do in certain situations: the people living on the 'hyperbolic' disk can achieve Euclidicity through resorting to Poincaré's temperature gradient. But to repeat: we are here talking about the logical and not about the heuristic or epistemological order of things.

Going back to the conjunction $G \wedge P_G$, we now see why any change in G might induce a change throughout a scientific system, thus allowing it to obviate an empirical refutation. For example: were P_G to state that light travels along geodesics, then it would trivially follow that P_G depends on the underlying geometry, G. Should we be faced with an empirical refutation, then we could decide to alter G into G'; for $G' \wedge P_{G'}$ generally fails to be observationally equivalent to $G \wedge P_G$ and might thus evade the refutation. We could alternatively change P_G into P'_G in such a way that: $\text{Not}((G \wedge P_G) \sim (G \wedge P'_G))$, where '~' stands for (strict) observational indiscernibility. We could for instance assume that light is deviated from a geodesic of G by some universal field $\underline{F}$ whose addition alters P_G into $P'_{G'}$.

The above considerations show that geometry functions both as a language and as a background theory permeating the whole of physics. Wherever $(G \wedge P_G) \sim (G' \wedge P'_{G'})$ holds good, Poincaré spoke—inappropriately—of the intertranslatability of the two geometries G and G'. It should by now have become clear that this misnomer actually refers to the observational identity of the two global systems $(G \wedge P_G)$ and $(G' \wedge P'_{G'})$.

Let us examine in finer detail Poincaré's contention that, given any G, $P_{G'}$ and G', then there always exists some $P'_{G'}$ such that $(G \wedge P_G) \sim (G' \wedge P'_{G'})$. As remarked above, we can arbitrarily dictate any geometrical structure we please, not because space as such is amorphous, but because it remains in principle inaccessible to our senses. Hence neither space itself nor any relations between various portions of it, nor the connection between space and any physical entity, will ever be observed. Only the direct connections between concrete objects like B_1 and B_2 could conceivably be ascertained. Let us recall that all relations between the ideal entities B°_1 and B°_2 are fixed by the geometry G. As for the connections between B_j and B^0_j (j = 1,2), they are determined by the physical theory which we have denoted by $P_{G'}$. G' will generally define a new ideal figure $B^{0'}_j$ corresponding to B_j; B^0_j and $B^{0'}_j$ may for example be geodesics of G and G' respectively (j = 1,2). Very schematically, it can be said that $P'_{G'}$ is obtained by adding to P_G the universal field $\underline{F}$ which effects the transition from B^0_j to $B^{0'}_j$ (j = 1,2). Thus:

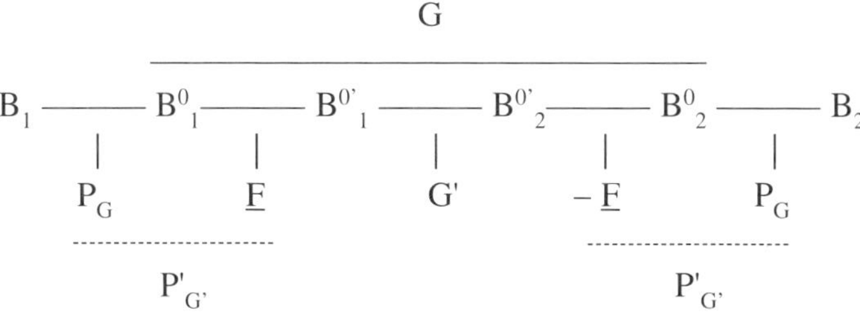

FIGURE 7

In the above scheme, the net relation between B_1 and B_2 remains unchanged; hence: $(G \wedge P_G) \sim (G' \wedge P'_{G'})$.

Grünbaum correctly points out that the introduction of the universal field $\underline{F}$ is tantamount to remetrizing space and thereby altering G into G'. He concludes that all talk of universal forces is purely metaphorical and that $\underline{F}$ can always be made to vanish by resorting to the 'customary means' of determining the $g_{ij}(\underline{x})$'s; on Poincaré's disk, for example, we could use material rods in order to measure distances and should thus be able empirically to construct a hyperbolic geometry without any mention of universal fields (Grünbaum 1973, pp. 86–87). This approach however faces two major difficulties: one of vicious circularity, the other of overdetermination.

Let us first note that Poincaré's disk example is over-idealized. Even though the temperature gradient be so contrived as to constitute a universal field, all measuring rods have to be corrected for the effects of differential forces, such as electromagnetic variations. Grünbaum rightly remarks that allowances invariably have to be made for certain inhomogeneous factors. But the latter are assessed through laws which in turn presuppose some kinematical framework G; and there is no guarantee that G will coincide with the geometry G' obtained by measurements made with the duly 'corrected' rods. Grünbaum proposes a most interesting iterative method which *might* yield some geometry G_0 as a fixed point of this iteration; that is, if all goes well, the geometry G_0 used in correcting the rods' lengths turns out to be identical with the G'_0 determined by means of these rods. (Grünbaum 1973, pp. 144–46). But again, there is no guarantee for the success of this operation.

For the sake of argument, assume that we have—in one way or another—succeeded in making allowances for all inhomogeneous factors. So let us transport our measuring rod to some distant point $P(\underline{x})$; we fix one end of the rod at P while letting its other extremity occupy a variable position denoted by some continuous index α. Let $ds_\alpha(\underline{x})$ be the length ascribed to the rod as it stretches from $P(\underline{x})$ to Q_α $(\underline{x} + d_\alpha\underline{x})$, where: $\underline{x} = (x^1,\ldots,x^N)$, $d_\alpha\underline{x} = (d_\alpha x^1,\ldots,d_\alpha x^N)$. Let us agree not to apply Einstein's summation convention to any Greek indices; so that $(ds_\alpha(\underline{x}))^2$ can be written as: $(ds_\alpha(\underline{x}))^2 = g_{ij}(\underline{x}) \cdot d_\alpha x^i \cdot d_\alpha x^j$. Since α is a continuous label, there are $\exp(2, \aleph_0)$ independent equations but only the $N(N+1)/2$ unknowns $g_{ij}(\underline{x})$ to be accounted for at any point P ($\aleph_0$ = aleph – zero). We consequently have no guarantee that the above equations will prove compatible. The requirement that our space be Riemannian *and* that its metric be realized by certain customary means is far from being innocuous: for we are imposing physical restrictions on the behaviour of the measuring rods and are thus no longer dealing with a mere stipulation, but with a genuine scientific hypothesis.

The diehard realist proceeds in the direction opposite to that of the conventionalist. Let us show that he will consequently face fewer epistemological problems. He generally *starts* by putting forward a conjecture about the intrinsic metric of physical space—about the $g_{ij}(\underline{x})$'s—thus automatically determining the geometry $G = [g_{ij}(\underline{x})]$. He might then wonder whether some well-known measuring instruments, such as a set of identical rods, realize G. Suppose that they do not. Would there then be a fact of the matter to be explained? The conventionalist must give a negative answer to this question; for according to him, the g_{ij}'s arise from a stipulation which *creates* the metric structure of space. After choosing the g_{ij}'s, he simply takes cognizance of a fact which has to do as much with the physical world as with his freely chosen convention, namely that the geometry realized by certain customary means—in case it is at all realizable in this way—does (or does not) coincide with the one he has arbitrarily stipulated. If it does, then well and good. If it does not, then there is still nothing to be explained; for according to conventionalism, we are at liberty to change our minds and remetrize space in conformity with some other set of measuring devices. The realist answers the same question very differently from the conventionalist. Assume that our rod attributes different distances to two infinitesimal stretches $\underline{PQ} = d\underline{x}$ and $\underline{PK} = \delta\underline{x}$, where: $\underline{x} = (x^1,\ldots,x^N)$, $\underline{x} + d\underline{x} = (x^1 + dx^1,\ldots,x^N + dx^N)$, $\underline{x} + \delta\underline{x} = (x^1+\delta x^1,\ldots,x^N + \delta x^N)$ are the coordinates of P, Q, K respectively. Suppose further that $g_{ij}(\underline{x}) \cdot dx^i \cdot dx^j = g_{ij}(\underline{x}) \cdot \delta x^i \cdot \delta x^j$. The realist will conclude that the rod is subject to distorting forces dependent on the directions $\underline{PQ}$ and $\underline{PK}$. Similarly: let two intervals $\underline{PQ}$ and $\underline{UV}$, which are distant from each other, be given different lengths by the rod, while $g_{ij}(\underline{x}) \cdot dx^i \cdot dx^j = g_{ij}(\underline{z}) \cdot dz^i \cdot dz^j$ (or the other way round); where $(\underline{z})$ and $(\underline{z} + d\underline{z})$ are the coordinates of U and V respectively. The realist might conjecture that, during its transport from P to U, the rod was deformed by a field whose magnitude and direction can in principle be discovered. He might alternatively conclude that his initial conjecture about the structure of space was erroneous; he would then try to alter the values of the $g_{ij}(\underline{x})$'s or even resort to an affine non-Riemannian structure.

At first sight, it might look as though the differences between conventionalism and realism were of a purely verbal nature: what for the conventionalist is a change of stipulation turns out, for the realist, to be the refutation of a hypothesis followed by the proposal of an alternative geometry. There remains however an important difference between these two methodologies; a difference which, to my mind, is all to the advantage of realism.

Grünbaum rightly claims that conventionalism explains a *fact* which realism fails to account for, namely the observational indiscernibility of any two physical geometries: according to conventionalism, space is metrically amorphous; so no wonder why no experiment can tell two arbitrarily chosen geometries apart (Grünbaum 1973, pp. 29–30).

The main problem however lies with the status of this presumed *fact.* My claim is that this fact has a logical rather than a strictly empirical character. For any proposition Y, let us denote by C[Y] the consequence class of Y, that is: C[Y] is the set of all logical consequences of Y. Put $Z = G \wedge P_G$ and $Z' = G' \wedge P'_{G'}$ (see above, p. 000). We shall assume that at any given time there is a consensus as to what constitutes an observation statement O, irrespective of whether O is true or false. Let E be the class of all such O's. We can take Poincaré to have established that $C[Z] \cap E = C[Z'] \cap E$; so there can be no crucial experiment between Z and Z' or—*a fortiori*—between G and G'. The last equation however expresses *no empirical fact.* Though membership of E be in part determined by our perceptual apparatus, no experimental findings—which would of course have to be described by certain elements of E—can either fasify or confirm $C[Z] \cap E = C[Z'] \cap E$; for this equation holds independently of the truth-values of any members of E. Hence the thesis of the amorphousness of space must be either logical or else metaphysical in Popper's sense, which is in no way meant to be a criticism of this thesis. However, the equation $C[Z] \cap E = C[Z'] \cap E$ acquires philosophical significance only in the light of a positivist *metaphysical* principle; namely that where there exists no detectable difference, there ought to be no objective difference either. Only then are we entitled to claim that since G cannot be *empirically* distinguished from G', space must lack a metric and the difference between G and G' be regarded as linguistic. Such a positivist thesis however remains highly questionable.

Needless to say, diehard realism is equally metaphysical, for it maintains that there exists an intrinsic spatial metric which, when properly captured, gives rise to a unified, organically compact, and hence 'convenient' system; where it should again be remarked that the convenience in question is not of the pragmatic kind. There is no denying that since it attributes an objective status to varying degrees of convenience, this thesis has a theological aspect. Tthis kind of theology is however perfectly compatible with thoroughgoing atheism, for it maintains that God-or-Nature combines the greatest variety of the low-level phenomena with the maximum unity and paucity of the underlying fundamental principles. (Leibniz pointed out that some optimal balance is thereby presupposed.)

I shall now attempt to tip the balance slightly in favour of structural realism. As correctly pointed out by Grünbaum, gravitation constitutes no genuinely universal field. It has nonetheless been conceded that throughout large domains of space-time, Einstein's General Theory of Relativity (GTR) can be reformulated in 'flat' terms. Hence classical and Einsteinian kinematics cannot be empirically told apart. Yet scientists rightly feel that the use of a Euclidean metric in GTR is artificial, and that such an artefact destroys the close coupling between gravitation and the space-time metric. Thus they hold that more than a merely linguistic difference separates these two formulations of General Relativity; and their reasons have nothing to do with whether or not gravitation is a genuinely universal force. They take kinematics to reflect an important aspect of the *real* world. This defence of geometrical realism clearly rests on metaphysical assumptions and is therefore far from decisive. It should nonetheless be added that by the very nature of the addressed problems, all rival solutions are bound to be metaphysical. Let me now show, by means of a rational reconstruction, that diehard realism can be legitimately attributed to Poincaré himself.

To repeat, Poincaré's thesis about the absence of an *observable* metric is not ontological but epistemological: human beings happen to be incapable of perceiving empty space. He moreover introduced the notion—as distinct from the terminology—of *universal fields*; and this not in an exclusively metaphorical sense; or rather: in no more of a metaphorical sense than that in which he used other theoretical terms like 'force', 'ether' or 'electron'. It is finally worth emphasizing that he took some metaphors to possess strong objective import.

Let us note that in the conjunction $G \wedge P_G$ the component P_G might already postulate a universal field; so that the remetrization of space issuing in the equivalence $(G' \wedge P'_{G'}) \sim (G \wedge P_G)$, that is the introduction of another universal force, could well cancel the one present in P_G. The new physics $P'_{G'}$ would then contain no universal fields at all; so that the new geometry G' might be the one based on what Grünbaum calls the customary definition of congruence. Be it as it may, since Poincaré made a purely methodological point about the arbitrariness of the metric, he was left free to make a further claim: if G renders the global scientific system $Z = (G \wedge P_G)$ maximally convenient and if Z either predicts or else systematically subsumes a whole host of facts, then such a high degree of convenience should not be attributed to mere chance. In Duhem's words: Z can then be regarded as approximating to a *Natural Classification*, so that any G fully integrated into Z must also reflect true relations between real entities. Such a conclusion could not possibly be drawn by Reichenbach for whom G is a mere *mode of description* of the facts of nature. Poincaré wrote:

> It will be said that science is a classification and that a classification cannot be true but convenient. But it is true that it is convenient; it is true that it is so not only for me but for all men; it is true that it will remain convenient for our descendants; it is true finally that *this cannot be by chance*. (Poincaré 1906a, p. 140; my italics)

We now see how close Poincaré's position comes to that of the diehard *structural* realist. To repeat: according to the latter, space possesses an intrinsic metric and hence a geometry which the scientist sets out to capture. The realist will thus feel entitled to hold that a geometry G will be a good approximation to the truth if G provides a coherent kinematical framework for a unified system Z, where Z systematically subsumes some facts and successfully anticipates others. Given his epistemological truth-concept, Poincaré would of course deny that G or, for that matter, any isolated convention can legitimately possess a truth-value. His methodological verdict nevertheless parallels that of the realist: Poincaré accepts that Z, alongside its various components and the way in which they cohere with each other, mirrors real relations obtaining in a transcendent world. This appraisal can moreover be transferred to the geometry G which subtends Z. As explained above, such a verdict remains unavailable to the conventionalist who maintains that space is amorphous and that geometry is decreed by the physicist; from which it immediately follows that there exists no structure which geometry could reflect rather than impose.

As repeatedly mentioned, Poincaré's pronouncements on the debate between realism and conventionalism were not always consistent. But the whole tenor of his philosophy is that convenience is not a merely subjective feature of an empirically successful system. On the contrary: convenience is the hallmark of high verisimilitude; and the degree of convenience of a geometrical theory is no exception to this rule of linking convenience to truth-likeness.

Despite this undeniable parallel between Poincaré's and the realist's methodological stands, the latter remains to my mind more coherent than the former. By holding that all statements are truth-bearers, the realist can justify his consistent use of a logic whose main function is to transmit truth downwards. A conventionalist à la Duhem, or even à la Poincaré, cannot offer any such a justification. Duhem was anyway aware of this gap in his position: he claimed that by divine decree, our *minds* must always obey the rules of traditional logic (Duhem 1914, Chapter 4, §9). We shall see that this Duhemian thesis is undermined by Poincaré's quasi-intuitionistic views concerning the foundations of mathematics (see below, Chapter 5).

4

Poincaré's Relativity Programme

To my knowledge, Edmund Whittaker is the only historian to have attributed the discovery of Special Relativity (STR) exclusively to Lorentz and Poincaré. He does this in a chapter of his *History of the Theories of Aether and Electricity*, in which Einstein is barely mentioned. Though unjust towards Einstein, Whittaker's *positive* account of Poincaré's achievement contains much more than a grain of truth (Whittaker 1953, Chapter 2). In what follows I propose to defend a thesis as forthright as Whittaker's, namely the view that Poincaré did discover STR, also that his structural realism and his 'conventionalist' epistemology provided him with essential heuristic guidelines, where, as shown in the previous chapter, the word 'conventionalism' should be regarded as a misnomer. This account in no way tarnishes Einstein's merit: following a path parallel to but independent of Poincaré's, Einstein finally transcended STR by constructing a generally covariant theory of gravitation.

Going back to Poincaré: we shall see that in 1900 he had already given the operational definition of clock-synchronisation which has since been attributed exclusively to Einstein; and that by 1905, he had gone far beyond the theoretical results obtained by all the other Relativists. Long before Minkowski appeared on the scene, Poincaré had determined the structure of the Lorentz-group and founded his Relativity Programme on the notion of Lorentz-covariance in a 4-dimensional manifold. He was the first physicist to have consciously based his heuristics on a symmetry requirement conceived as a condition of form-invariance under a given set of transformations. The latter was used in order to correct a major error in Lorentz's law for the electric density and to find the transformation equations for the electromagnetic field, which is why it seemed appropriate that I should devote a whole section to a general study of the problems posed by symmetry principles. Poincaré furthermore reflected on the parallel between gravity and inertia as well as on the possibility of a dependence of the gravitational attraction on the velocity of a particle. He proposed a Lorentz-covariant theory of gravitation which yields Newton's inverse-square law as a limiting case for small speeds. He had thus made use not only of one specific symmetry requirement but also of the general Principle of Correspondence, which was later attributed to Niels Bohr. Given the heuristic importance of this principle, another section will be devoted to examining some of its presuppositions, more especially the assumption of the uniform continuity of certain functions. Poincaré had thus pushed STR to the utmost limits of which it seemed capable.

With regard to Lorentz, Poincaré, and Einstein, historians therefore have an important duty; namely that of not only acknowledging Einstein's unsurpassed contributions to Relativity, but also of rectifying a certain imbalance. After Poincaré's death in 1912, Einstein saw the necessity of going beyond STR towards a generally covariant theory of gravitation. Because of the great beauty of the general theory, there has been an unfortunate tendency to forget Poincaré's achievement and to regard Einstein as the sole founder of the Relativity programme, not only in its general, but also in its special phase. The present chapter is, in part, intended to redress this injustice.

4.1 First Steps towards Relativity

With regard to conventions, Poincaré put forward two theses and made two prognoses. He maintained that we shall always be in a position to hold on to Classical Mechanics *and* to Euclidean Geometry, provided we be prepared to alter—perhaps in a drastic way—the rest of physics. We have already seen that these two claims, and more particularly the second one, have a logical character and must consequently be considered valid, but Poincaré also predicted that Classical Mechanics *and* Euclidean Geometry would always turn out to be the most convenient systems and therefore have nothing to fear from fresh experiments (see above, Chapters 1 and 3). It is ironical that through criticising Lorentz's hypotheses and then launching the Relativity Programme, Poincaré was destined to refute the first prognosis. He unfortunately did not live long enough to accept—as he most certainly would have done—Einstein's refutation of his second prediction.

Let me present a survey of the most important results achieved by Lorentz and then reconstruct Poincaré's views from his criticisms of Lorentz's position. In the absolute or ether frame, the Maxwell-Lorentz equations can be written in the form:

(1) $\nabla \cdot \underline{E} = \rho$; $\nabla \cdot \underline{H} = 0$; $\nabla \times \underline{E} = -[\partial \underline{H}/\partial t]/c$; $\nabla \times \underline{H} = [(\partial \underline{E}/\partial t) + \rho \underline{v}]/c$;

where: ∇ denotes the operator $(\partial/\partial x, \partial/\partial y, \partial/\partial z)$, $\underline{E}$ the electric field, ρ the charge density, $\underline{H}$ the magnetic field, and $\underline{v}$ the charge velocity.

In what follows, we shall write:

(2) $\underline{E} = (E_1, E_2, E_3)$ and $\underline{H} = (H_1, H_2, H_3)$;

that is: E_j and H_j are the jth components of $\underline{E}$ and $\underline{H}$ respectively.

To these equations Lorentz added the expression of the poderomotive force $\underline{L}$ which acts on a particle moving with velocity $\underline{w}$ and carrying charge e:

(3) $\underline{L} = e[\underline{E} + (\underline{w}/c) \times \underline{H}]$.

In empty space (where $\rho = 0$), equations (1) reduce to:

(4) $\nabla \cdot \underline{E} = 0 = \nabla \cdot \underline{H}$; $\nabla \times \underline{E} = -[\partial \underline{H}/\partial t]/c$; $\nabla \times \underline{H} = [\partial \underline{E}/\partial t]/c$.

It can be shown that in any system to which the relations (4) apply, light and all other electromagnetic disturbances are propagated in all directions with the constant speed, c. Since the earth is in motion through the ether, Lorentz considered a Galilean frame G which moves with velocity $\underline{u}$ in absolute space. Without loss of generality, we can choose our x-axis to be parallel to $\underline{u}$. The following equations give the Galilean coordinates x_g, y_g, z_g, and t_g of an event in G in terms of its absolute spatio-temporal position (x, y, z, t):

(5) $x_g = x - ut$; $y_g = y$; $z_g = z$; $t_g = t$.

Note that the last equation, namely $t_g = t$, was hardly ever explicitly written down, for *time* was implicitly taken to be frame-independent, or in other words, absolute.

It is easily verified that Maxwell's equations are not Galileo-covariant: they do not keep the form (1) in all Galilean frames. This can be immediately inferred from the fact that, in G, the velocity of a light beam along the x-axis should be equal, not to c, but either to c-u or to c+u; it is therefore a variable quantity, so that equations (4) cannot possibly hold in G.

Partly in response to the null outcome of the Michelson-Morley experiment, Lorentz pursued two new lines of research. The first, which was certainly adhoc, will be discussed below. The second eventually yielded an explanation of Michelson's results through the introduction of the so-called *effective* coordinates: x", y", z", and t", which are defined as follows:

(6) $x'' = \gamma x_g = \gamma(x - ut)$; $y'' = y_g = y$; $z'' = z_g = z$; $t'' = \gamma[(t_g/\gamma^2) - ux_g/c^2] = \gamma(t - ux/c^2)$;

where: γ is defined by: $\gamma = 1/\sqrt{(1 - u^2/c^2)}$. (For more details, see Zahar 1989, Chapter 2.)

These equations constitute the Lorentz-Transformation, or L.T. for short. The inverse of (6) is obtained first by substituting –u for u, then by interchanging x and x', y and y', and so on. This yields:

(7) $x = \gamma(x'' + ut''); y = y''; z = z''; t = \gamma(t'' + ux''/c^2)$.

Lorentz was initially unaware of the fact that the spatio-temporal distance $(ds^{2-}-c^2dt^2)$ is an invariant of the L.T.; that is: that as a result of (6), we have:

(8) $dx^2 + dy^2 + dz^2 - c^2dt^2 = dx''^2 + dy''^2 + dz''^2 - c^2dt''^2$.

Over and above introducing the effective coordinates, Lorentz proposed, in 1904, the transformation of the field $<\underline{E}, \underline{H}>$ into $<\underline{E}'', \underline{H}''> = <(E_1'',E_2'',E_3''), (H_1'',H_2'',H_3'')>$ which is determined by:

(9) $E_1'' = E_1; E_2'' = \gamma(E_2 - uH_3/c); E_3'' = \gamma(E_3 + uH_2/c); H_1'' = H_1; H_2'' = \gamma(H_2 + uE_3/c)$; and $H_3'' = \gamma(H_3 - uE_2/c)$.

Since he gave no realistic interpretation of the *local time* t", Lorentz could ascribe no importance to a fundamental corollary of (8); namely that a process has velocity c in the stationary frame iff its *effective* velocity in G is also equal to c. Yet he knew that relations (4) keep their form under transformations (6) and (9); that is: that Maxwell's equations for the vacuum are Lorentz-covariant. Thus, if (4) holds, then we also have:

(10) $\nabla'' \cdot \underline{E}'' = 0 = \nabla'' \cdot \underline{H}''; \nabla''\times\underline{E}'' = -[\partial\underline{H}''/\partial t'']/c; \nabla''\times\underline{H}'' = [\partial\underline{E}''/\partial t'']/c;$

where we define: $\nabla'' = (\partial/\partial x'', \partial/\partial y'', \partial/\partial z'')$.

From these results, Lorentz concluded that no uniform translation in the ether gives rise to any first-order observable effects (first-order, that is, in (u/c)). STR can be looked upon as a generalisation of this so-called theorem of corresponding states, and Lorentz himself took a first step towards this generalisation by showing that in the case of Michelson's experiment, the contraction hypothesis explains the absence of all the second-order optical effects which had been attributed to the earth's motiopn (Lorentz 1904).

No doubt because of his commitment to the concept of absolute time, Lorentz either never managed or else was reluctant to interpret t" realistically. In most applications, he maintained that the difference between t and t" was so negligible as to make these two quantities practically equal. In fact, not only did he omit to give any physical interpretation of the *local time* t", he furthermore continued to regard the Galilean transformation (5) as the only legitimate one. He called the quantities x", y", z", t" *effective coordinates*; that is: he regarded them as mere instruments for carrying out certain caculations and for making empirical predictions. This eventually prevented him from correctly interpreting the transform ρ" of the charge density ρ and hence from establishing the full Lorentz-covariance of the Maxwell-Lorentz equations (1) (for more details, see Zahar 1989, Chapter 3).

Let us now turn to Lorentz's adhoc solution, namely to the contraction hypothesis as presented in one section of his 1895 *Versuch* (see Einstein et al. 1923, Section 1). The reasons for dwelling on this adhoc theory—or rather, on this whole spectrum of adhoc conjectures—are twofold. First, such an account provides a *rationale* for Poincaré's remark that the contraction hypothesis was doctored post-hoc in order to explain away Michelson's bothersome results. We shall secondly see that Lorentz's adhoc law provided a starting point for Poincaré's own work on the Relativity Programme.

We shall provisionally accept the absolute space and time hypotheses, and we tentatively assume that a body moving through the ether will be contracted both along and normally to the direction of its absolute motion. Let φ and ψ respectively denote the longitudinal and transverse contraction coefficients affecting any mobile. Lorentz initially refused to speculate about the magnitudes of φ and ψ, which he nonetheless proposed to determine experimentally. Let x', y', and z' be the coordinates measured in a frame moving with a material body, where the x- and x'-axes are chosen to be parallel to the direction of motion. Thus:

(11) $x' = x_g/\varphi = (x-ut)/\varphi;\ y' = y_g/\psi = y/\psi;\ z' = z_g/\psi = z/\psi;$

where u is the speed of the frame.

Lorentz later realised that since all clocks are made of material rods, they will also be affected by their displacement in the ether (see Lorentz 1909, Notes). We shall denote by Ω the coefficient of clock-retardation, without deciding a priori whether or not $\Omega \neq 1$. Let Q and Q' be the stationary and mobile frames respectively. In Q consider two simultaneous events, F_1 and F_2, which take place on the same plane $x = x_0$. Hence F_1 and F_2 are of the form: $F_1 = (x_0,y_1,z_1,t_0)$ and $F_2 = (x_0,y_2,z_2,t_0)$. F_1 and F_2 must clearly be simultaneous in Q'; for otherwise, the times of occurrence of these events in Q' would depend on their y- and z-coordinates, which is inadmissible since both the origin O' of Q' and the directions of the y'- and z'-axes can be chosen arbitrarily. In other words, we could rotate and translate these axes without altering any objective state-of-affairs or any chosen clock-synchronisation convention. (These considerations do not apply to the choice of the x-axis whose direction is fixed by the velocity $\underline{u}$ of Q' with respect to Q.) Since F_1 and F_2 are simultaneous in Q', t_0' will depend exclusively on t_0 and x_0. Assuming our transformation to be linear, we can therefore write:

(12) $t' = At + Bx.$

Note that we have so far taken no account of the retardation of moving clocks by the factor Ω. In order to do so, let us consider a point spatially fixed in Q' and then write $dt' = \Omega.dt$. Using (11) and (12), we can express t' in terms of t and x'. Thus: $t' = (a + Bu)t + B\varphi x'$. Since x' is taken to be constant, $dt' = (A + bu) \cdot dt$. Hence: $A + Bu = \Omega$. That is: $A = \Omega - Bu$. Substituting in (11) and (12), we obtain the general transformation:

(13) $x' = (x - ut)/\varphi$; $y' = y/\psi$; $z' = z/\psi$; $t' = (\Omega - Bu)t + Bx.$

These equations involve only one open parameter, namely B. We shall in fact show that B reflects a synchronisation procedure which can be chosen arbitrarily. It turns out that B is uniquely determined by the Poincaré-Einstein convention for clock-synchronisation. We can moreover immediately check that the Galilean transformation constitutes a special case of (13): in a classical world, there is neither clock-retardation nor length-contraction: $\varphi = \psi = \Omega = 1$; taking B to be 0, we obtain the usual Galilean coordinates x_g, y_g, z_g, and t_g.

4.2 Convention and Physical Fact

Poincaré remarked that all observed motions are relative and that no movement with respect to the ether can be directly ascertained. This *crude fact* instinctively leads us to subscribe to a universal principle of relativity, that is: to a belief prior to any articulated hypothesis. Concerning the ether, Poincaré is prepared to regard the assumption of its 'existence' as a convention but he reminds us that the latter remains acceptable only as long as it proves convenient. He then formulates his restricted principle of Relativity as the requirement that all natural laws assume the same form in all inertial frames. This is in fact Einstein's formulation, but in *Science and Hypothesis*, Poincaré linked Relativity to the Galilean group while adding that because of our general adherence to this principle, we would have wished to extend it to all frames without exception. We are however prevented from doing so by the crude fact that accelerated motions give rise to detectable effects.

With regard to the concept of time, Poincaré maintains that there exists only one crude fact which cannot be ignored; namely the fixed temporal order obtaining within *each individual* consciousness. We have however no intuitive idea of the numerical equality of two durations or of the simultaneity of two distant events, which either are spatially separated or else take place in the minds of two distinct individuals. Thus physical time presupposes a *convention*, where the latter should—as usual—be appraised exclusively in terms of the convenience of the hypotheses in which it is used. Now we understand why, not being a priori committed to the notion of absolute time, Poincaré found it easy to abandon it in favour of the more convenient concept of local time.

In 1895 he maintained that despite being far superior to its rivals, Lorentz's hypothesis violates the principles both of Relativity and of reaction. We have already seen that since they are not Galileo-covariant, the Maxwell-Lorentz equations contradict the classical principle of Relativity. Towards the end of his paper: 'À propos de la théorie de M. Larmor', Poincaré adds a remark which, though seemingly important, is left unclarified, namely that the violation of Galilean invariance is closely connected to that of Newton's Third Law—the principle asserting the equality of action and reaction (Poincaré 1895).

In 1900-01 Poincaré offered a tentative solution to this problem. He knew that Lorentz had used an uninterpreted notion of local time in order to explain the absence of all the first-order efects which had been attributed to the earth's motion (see Poincaré 1900 and 1901). As already explained, his position enjoined him to regard all scientific theories as testable by means of crude facts. The observations yielding such facts generally involve time measurements, which are made possible by certain stipulations. We have seen that only the local time t" plays any operative role in the testable predictions derived from Lorentz's theory. Poincaré was thus led to give an operational definition of t" through a convention which was later attributed exclusively to Einstein. Two distant clocks H and K, which are at rest relative to each other, are said to be synchronous iff: having left H at H-time t, a light ray reaches K at K-time $t + t_0$, is then instantaneously reflected at K and goes back to H, where it arrives at H-time $t + 2t_0$. Should these relations hold good for all values of t, then light will, by convention, have the same speed from H to K as from K to H; but there is no guarantee that the velocity of light must also be equal to c (= 300000 kms/s). Be it as it may, the time thus determined by synchronous clocks turns out to be identical with Lorentz's local time. Simultaneity will thus depend on the state of motion of the composite system HK, that is: on the chosen frame of reference.

Now take H and K to be two points of the stationary frame Q which have the fixed spatial coordinates (x, 0, 0) and (x + h, 0, 0) respectively. Assuming that h>0, we infer from Maxwell's equations that:

(14) $(h/t_0) = c$ = velocity of light.

In order to calculate the speed of any signals between H and K, we go back to (13) which entails: $\delta x' = (\delta x - u \cdot \delta t)/\varphi$; $\delta t' = (\Omega - B \cdot u)\delta t + B \cdot \delta x$. Hence:

(15) $v' =_{\text{Def.}} \delta x'/\delta t' = [(\delta x/\delta t) - u]/[\varphi((\Omega - Bu) + B \cdot \delta x/\delta t)] = [c - u]/[\varphi((\Omega - Bu) + Bc)]$;

where v' is the velocity—in the moving frame Q'—of the light signal from H to K; for note that between H and K: $\delta x = h$, $\delta t = t_0$ and $h/t_0 = c$. For the signal from K to H, we have to take $\delta x = -h$ and $\delta t = t_0$, thus obtaining, in view of $h/t_0 = c$:

(16) $w' = [-c - u]/[\varphi((\Omega - Bu) - Bc)]$

where w' is the velocity—in Q'—of the light signal from K to H.

Let us now adopt the Poincaré-Einstein synchronisation convention which implies that $v' = -w'$ = round-trip velocity along HKH. By (15) and (16): $[c - u]/[\varphi((\Omega - Bu) + Bc)] = [c + u]/[\varphi((\Omega - Bu) - Bc)]$. Solving for B, we obtain:

(17) $B = -u\Omega/[c^2 - u^2] = -\Omega u\gamma^2/c^2$;

where, as usual: $\gamma = 1/\sqrt{(1 - u^2/c^2)}$.

This specific value of B is therefore fixed by the synchronisation procedure introduced by Poincaré. Substituting back in (13), we have the following equations:

(18) $x' = (x - ut)/\varphi$; $y' = y/\psi$; $z' = z/\psi$; $t' = \Omega\gamma^2(t - ux/c^2)$.

As already explained, this is the most general linear transformatin which incorporates the Poincaré-Einstein convention. Thus it ensures that in Q', the velocity of light from any point H to any other point K is the same as that from K to H; but this velocity will generally differ from c.

Let us actually determine the speed of light as measured in the moving frame Q'. Putting: $x^* = \gamma\varphi x'$, $y^* = \psi y'$, $z^* = \psi z'$ and $t^* = t'/(\Omega\gamma)$, we infer from (18):

(19) $x^* = \gamma(x - ut); y^* = y; z^* = z; t^* = \gamma(t - ux/c^2)$.

In other words: (x^*, y^*, z^*, t^*) are obtained from (x, y, z, t) by means of an ordinary Lorentz-Transformation (LT) (see (6) above). It follows from (8) that:

(20) $dx^2 + dy^2 + dz^2 - c^2 \cdot dt^2 = dx^{*2} + dy^{*2} + dz^{*2} - c^2 \cdot dt^{*2} = [\gamma^2 \cdot \varphi^2 \cdot dx'^2 + \psi^2(dy'^2 + dz'^2) - c^2 \cdot dt'^2/(\Omega^2\gamma^2)]$, by the definition of (x^*, y^*, z^*, t^*).

Consider any process **P** which unfolds with speed c in Q. The first member of (20) will therefore vanish. Let $\underline{c}' = (c'_1, c'_2, c'_3)$ be the velocity vector of **P** in Q'; that is: $c'_1 = dx'/dt'$, $c'_2 = dy'/dt'$, $c'_3 = dz'/dt'$. (20) implies that:

(21) $\gamma^2 \cdot \varphi^2 \cdot c'^2_1 + \psi^2(c'^2_2 + c'^2_3) - c^2/(\Omega^2\gamma^2) = 0$.

Putting $c' = |\underline{c}'|$ and denoting by α the angle between $\underline{c}'$ and the x'-axis of Q', we have: $c'^2_1 = c'^2 \cdot \cos^2\alpha$ and $c'^2_1 + c'^2_2 = c'^2 \cdot \sin^2\alpha$. Substituting in (21):

(22) $c'^2 = c^2/[\Omega^2\gamma^2(\gamma^2 \cdot \varphi^2 \cdot \cos^2\alpha + \psi^2 \cdot \sin^2\alpha)]$.

As anticipated, c' depends on the direction α along which the process **P** takes place.

Since the Poincaré-Einstein convention yields a nominalist definition of simultaneity, it can have no empirical content or should at any rate prove neutral as between rival hypotheses. It could admitedly dictate that the round trip and the one-way velocity of light are the same along every direction in Q'; but the actual magnitude of the round-trip speed represents a physical datum which cannot—by *fiat*—be required to be independent of α. Nothing can consequently prevent us from adopting the Poincaré-Einstein convention in *any* world, more particularly in the classical one where: $\varphi = \psi = \Omega = 1$. Thus, by (18) and (22): $x' = x - ut$; $y' = y$; $z' = z$; $t' = \gamma^2(t - ux/c^2)$; $c'^2 = c^2/[\gamma^2(\gamma^2 \cdot \cos^2\alpha + \sin^2\alpha)]$. It is hardly surprising that in a Newtonian universe, c' should vary with α; that is: that in Q', the speed of light should not be the same in all directions.

Let us now consider the Michelson-Morley experiment. One way of accounting for its null result is to postulate the independence of c' from α; in other words: to resort to an adhoc empirical conjecture. This is precisely what Lorentz did in one section of his *Versuch* (see Lorentz 1895). He assumed that $\gamma\varphi = \psi$. By (22) and (18), it follows that:

(23) $c'^2 = c^2/(\Omega\psi\gamma)^2$;

that is: $c' = c/(\Omega\psi\gamma)$; and:

(24) $x' = \gamma(x - ut)/\psi$; $y' = y/\psi$; $z' = z/\psi$; $t' = \Omega\gamma^2(t - ux/c^2)$.

These equations, which constitute the *adhoc* response of Lorentz to Michelson's results, provided Poincaré with the starting point for his work on STR. (We should recall that Lorentz proposed another solution to the problem posed by Micheson's experiment; namely the Molecular Forces hypothesis (MFH), which is far from being adhoc.)

As a consequence of (23), the speed of light c' in Q' will be the same in all directions. This explains why the interference fringes in Michelson's interferometer will not be displaced when the apparatus is rotated. But also by (23), we could still have: $c \neq c/(\Omega\psi\gamma) = c'$; that is: unless $\Omega\psi\gamma = 1$; or equivalently: $\Omega\psi = 1/\gamma = \sqrt{(1 - u^2/c^2)}$. This last assumption, which was made by both Einstein and Lorentz, is tantamount to the assertion that c is a fundamental invariant.

In conclusion, the choice of B reflects a mere synchronism *convention.* Let us note that transformation (13) is linear and hence one-one and well-behaved. We can therefore choose any stipulative definition of clock-synchronisation in a world in which clocks in motion are retarded by the factor Ω while material bodies are longitudinally and transversally contracted by the coefficients φ and ψ respectively; where Ω, φ, and ψ are arbitrary well-behaved functions of the speed of the mobile. Hence, contrary to popular belief, no novel or more precise definition of simultaneity either entails or even leads to a new physical hypothesis like STR. It is only against the background of a *pregiven* theory that a convention proves to be more or less convenient. Poincaré's synchronisation procedure actually yields the simplest definition of coordinate time in an Einsteinian, but not in a Newtonian world. (For more details, see Zahar 1977.)

4.3 An Outline of Poincaré's Relativity Programme

Let us reconsider Poincaré's insight that Newton's Third Law involves a notion of simultaneity incompatible with Maxwellian electrodynamics. Let A_1 and A_2 be two electrons such that A_1 acts on A_2 across the distance k separating A_1 from A_2. After leaving A_1, an electromagnetic disturbance will be propagated towards A_2 with the speed of light c. A_1 will therefore have resumed its state of rest when, after time k/c, the disturbance reaches A_2. Hence the action on A_2 is *not instantaneously compensated* by a reaction, which can reach A_1 only after a second interval k/c. Thus Maxwell's equations contradict Newton's Third Law because the finite velocity of signal-propagation cannot be reconciled with the simultaneity of action and reaction. One moreover begins to understand how the Third Law connects with Galilean Relativity: the latter rests on the notion of absolute simultaneity, that is: on the invariance of the relation $\delta t = 0$, which is also presupposed by the principle of reaction. So it comes as no surprise that Lorentz's theory should both conflict with Galilean Relativity and transgress Newton's Third Law. According to Poincaré, we would no doubt have jettisoned the principle of relativity, had it proved irreconcilable with experience and with accepted background knowledge. But the 'crude' facts themselves, in the form of Michelson's results, had stubbornly confirmed the relativity of all optical phenomena. Poincaré accused Lorentz of having accounted for these results by resorting to a whole panoply of disparate hypotheses. This criticism seems to me unfair because it ignores Lorentz's non-adhoc solution of the Michelson-Morley problem. In an important section of the *Versuch,* Lorentz introduced only one auxiliary assumption, namely the MFH, according to which all forces are transmitted by the same ether and must therefore transform like the Lorentz-force. Being consonant with the ether-theoretic approach, the MFH is a plausible hypothesis which also unexpectedly entails the retardation of moving clocks and the contraction of measuring rods (see Zahar 1989, Chapter 2). Poincaré could justifiably have maintained, not that Lorentz had accumulated a whole host of hypotheses, but that he failed to connect the MFH with his new concept of local time. Lorentz was however unable to establish such a connection for he had omitted—or simply refused—to give a realistic interpretation of t''. This is precisely the gap in Lorentz's approach which Poincaré's synchronisation convention was meant to fill. But Poincaré expected from himself and from others far more than the stipulation of a mere convention. Already in his 1902, he had looked upon the absence of first and second-order optical effects, not as *accidents of Nature*, but as instances or illustrations of one universal principle. In conformity with his general philosophy and with the Relativity principle governing classical mechanics, he had *generalised* this principle by proposing that uniform rectilinear motion should give rise to no detectable effects whatsoever. He had thus elevated Relativity to the rank of a meta-postulate which applies, not only to mechanics, but also to elctromagnetism. At the 1904 Saint Louis Conference, he underscored the necessity of creating a unified system which would genuinely explain all the prima facie disparate electromagnetic phenomena; and it was he who subsequently (in 1905) constructed such a theory (Poincaré 1906a, pp. 109–110).

Let us recall that Poincaré judges a convention to be convenient if it is both unified and susceptible of accounting for experience without recourse to adhoc hypotheses; also that in Lorentz's system, all the observable results are derived via the *effective* coordinates; in other words: via quantities which, being effectively measured, describe either crude facts or else states-of-affairs not far removed from genuinely crude facts. The Galilean Transformation, though taken by Lorentz to be the only 'true' one, breaks the smooth transition from the 'absolute' to the 'effective' coordinates (see equations (5) and (6) above). We finally know that unlike Lorentz, Poincaré denied scientific hypotheses any *direct* ontological import; so he naturally proposed to abolish the Galilean Transformation and base Relativity on the 'effective' variables. In this way, he obtained, as a first heuristic component of his programme, the *Principle of Lorentz-covariance* which is in effect a *symmetry* requirement. As for the second component, namely the *Correspondence Principle*, he regarded it as an indispensable instrument to be used in all fields of scientific inquiry. Given the heuristic importance of these two principles for the development of modern physics, we shall discuss them from a general point of view rather than from the narrow perspective of STR.

4.4 The Correspondence Principle

The Correspondence Principle is of special significance for Poincaré's general philosophy of physics; for we have just seen that qua heuristic device, it forms an essential part of his Relativity Programme. Though attributed to Niels Bohr, this principle had long been invoked, for instance by Newton, who claimed to have 'derived' his theory from Kepler's laws of planetary motion. We should however remember that the latter constitute, not logical consequences, but limiting cases, of Newton's equations.

As regards Poincaré, what is remarkable is not only that he systematically made use of the Correspondence Principle in his Relativity Programme, but also that he discovered a phenomenon which seems to fly in the face of this principle, namely the phenomenon of chaos. Roughly speaking, this is the fact that no matter how close to each other two functions might initially be, they could eventually differ from each other in the most unexpected and spectacular manner. Consider Newton's and Einstein's laws of motion, the equations

$\underline{f} = d(m\underline{v})/dt$ and $\underline{f} = d[m\underline{v}/\sqrt{(1 - v^2/c^2)}]/dt$.

Though they provide no genuine instance of 'chaos', these two relations start by being observationally equivalent for 'small' values of (v/c); but in the neighbourhood of (v/c) = 1, they diverge qualitatively from each other in a most dramatic way. Yet the relativistic equation not only represents a limiting case of the classical one, but taken together with a requirement of Lorentz-covariance, it was actually *deduced* from a special case of the classical relation $\underline{f} = d(m\underline{v})/dt$ (see Zahar 1989, Chapter 7).

It is therefore essential for us to characterise the Correspondence Principle in a general way which ought to account for the twin aspects of continuity and of divergence in all scientific revolutions—a way which explains continuity while leaving room for revolutionary change. Through invoking the Heine-Borel theorem, it will be shown that a new *revolutionary* theory can—in certain respects and within prescribed domains—be *uniformly continuous* with the very hypothesis it is meant to supersede. This confirms Planck's claim that the best theory is normally the most *conservative* solution to a *revolutionary* problem (Planck 1965c).

Science is classically portrayed as evolving through a succession of improved approximations to the truth. Despite renouncing the goal of absolute certainty, Duhem held the sequence of all physical theories to be homing in on a Natural Classification which would reflect the order of things as present to God's mind (Duhem 1914, Chapter 2). Popper similarly looked upon science as having succeeded in producing a sequence of conjectures with increasing degrees of verisimilitude (Popper 1963, Chapter 10). Apart from the technical difficulties encountered by the formal definition of verisimilitude, this cumulative view of progress was challenged by the following observation: at the ontological level, the development of science displays an alarming degree of referential discontinuity: as we move from one hypothesis to the next, the same-sounding word is often found to denote, or rather to conflate, two fundamentally different types of entity. For example: the term 'electromagnetic waves' referred, first of all, to the mechanical states of the ether; then to two oscillating fields deprived of any material support; finally, to massless elementary particles called 'photons'.

The cumulative view can however be rescued by an appeal to two essential aspects of the development of empirical science; and these aspects turn out to be closely linked. First, it is generally agreed that at the practical level, our control over physical nature has steadily grown. This progress was secondly taken to have been achieved by the systematic application of a continuity requirement, namely the Correspondence Principle, which demands that old laws be either logical consequences or limiting cases of new hypotheses. This essentially mathematical form of continuity explains why a new theory inherits the degree of confirmation of its predecessors. Let us however note that identity is the most perfect form of continuity; but in the midst of a scientific revolution, identity is the last condition we would wish to impose. In well-specified areas, the new hypothesis is expressly meant to differ markedly from the old one. For example, a new law may be required to avoid, or rather to overcome the known refutations of previous conjectures.

This tension between continuity and saltation can be resolved by means of the notion of one scientific system being *imbedded in*—rather than forming *a substructure of*—another one (see Redhead 1995, Chapter 4). The following examples will clarify this important distinction. Even though the complex plane **C** extends and hence imbeds the real line **R,** the former does not mirror *all* the properties of the latter, *nor vice versa;* for the linear ordering of **R** cannot be naturally extended to **C** while **C,** but not **R**, is algebraically closed. Another example is provided by group theory: a non-commutative group G contains a whole family of cyclic subgroups H, each generated by some element $y \in G$. Thus every H is both Abelian and imbedded in the group G, but since the latter is not commutative, it does not reflect all the structural properties of H, nor vice versa. Finally: we know that hyperbolic geometry possesses a model in the complex plane **C** and can consequently be taken to be imbedded in **C**. It would however be absurd to claim that hyperbolic geometry, which is *non-Euclidean*, constitutes a substructure of the Euclidean plane.

Michael Redhead gives an important example of discontinuity taken from a chapter of relativistic physics (Redhead 1995, Chapter 4). Putting $\lambda = 1/c$, treating λ as a variable parameter and writing LT(λ) for the Lorentz-transformation, we can regard the Galilean Transformation GT = LT(0) as a special case of LT(λ). But $1/c \neq 0$, so that unlike LT(1/c), GT = LT(0) is singular. In other words: by continuity, we have: LT(λ) $\rightarrow$ GT as $\lambda \rightarrow 0$; yet, for all values of c, det(GT) = 0 $\neq$ det(LT(1/c)). Despite the fact that GT is a limiting case of LT(1/c), non-singularity can be looked upon as an 'emergent' property of LT(1/c). This property furthermore accounts for essential features of relativistic physics like the close connection between space and time, as well as between the two conservation laws of momentum and energy.

Remembering that one of my aims is to show the relevance of mathematics to the development of physics, I propose to demonstrate that the possibility of reconciling the gradualist with the revolutionary aspects of scientific progress rests on a topological result; namely on the Heine-Borel theorem which tells us that all closed and bounded subsets of $\mathbf{R}^n$ are compact, where n is an arbitrary positive integer. Consider the hypothesis:

(25) $(\forall v,a,b)[Q(v,a,b) \Rightarrow (h(v,a,b) = 0)]$;

which is taken to have been superseded by:

(26) $(\forall v,a,b)[Q(v,a,b) \Rightarrow (H(\lambda_0,v,a,b) = 0)]$;

where λ_0 denotes a small constant.

Replacing λ_0 by the variable parameter λ, we shall assume that $H(\lambda,v,a,b)$ and $h(v,a,b)$ are continuous functions of $<\lambda,v,a,b>$ and of $<v,a,b>$ respectively. Now suppose that $[h(v,a,b) = 0]$ has been strongly corroborated in some region Δ, that is: that this law has withstood a number of severe tests in Δ. This number being necessarily finite, we can assume, without any loss of generality, that Δ is of the form:

(27) $\Delta = V \times A \times B$, where V, A, and B are bounded closed sets of real numbers.

For example, we can take A to be the closed interval [a',a"] where a' and a" are respectively the minimum and the maximum measured values of a. Similar considerations apply to V and to B. When restricted to Δ, the laws (25) and (26) assume the forms:

(28) $(\forall v,a,b)[(Q(v,a,b) \wedge (<v,a,b> \in \Delta)) \Rightarrow (h(v,a,b) = 0)]$;

and:

(29) $(\forall v,a,b)[(Q(v,a,b) \wedge (<v,a,b> \in \Delta)) \Rightarrow (H(\lambda_0,v,a,b) = 0)]$, respectively.

In view of the approximate character of all measurements, we can take (28) to be observationally indiscernible from:

(30) $(\forall v,a,b)[(Q(v,a,b) \wedge (<v,a,b> \in \Delta)) \Rightarrow (|h(v,a,b)| \leq \mu_0)]$,

where μ_0 is some fixed positive real.

(30) is generally taken to hold not only for the values of v, a, and b which are actually measured, but also for all intermediate values; that is: throughout the domain Δ. In other words, (30) is regarded as true.

In one of its forms, the Correspondence Principle requires that:

(31) for any $<v,a,b> \in \Delta$, $H(\lambda,v,a,b) \to h(v,a,b)$, as $\lambda \to 0$.

Given the continuity of H, we also have: $H(\lambda,v,a,b) \to H(0,v,a,b)$, as $\lambda \to 0$.

By the uniqueness of the limit, we infer from (31) that:

(32) $h(v,a,b) = H(0,v,a,b)$, for all $<v, a, b> \in \Delta$.

Let us describe the way in which the new theory H accounts for the empirical success of h in Δ. We provisionally accept (29), from which (30) will now be deduced. Since λ_0 is small, there exists a real number $k > 0$ such that:

(33) $\lambda_0 \in [-k, +k]$.

Let the parameter λ be confined to the closed interval $[-k, k]$. By the continuity of H and the compactness of $[-k,k] \times V \times A \times B = [-k,k] \times \Delta$, the following theorem about uniform continuity holds good. For any $\mu > 0$, there exists an $\eta(\mu)$, dependent only on μ, such that:

$|H(\lambda_1,v_1,a_1,b_1) — H(\lambda_2,v_2,a_2,b_2)| \leq \mu$, for all $<\lambda_j,v_j,a_j,b_j> \in [-k,k] \times \Delta$ ($j = 1,2$) satisfying: $\sqrt{[(\lambda_1 - \lambda_2)^2 + (v_1 - v_2)^2 + (a_1 - a_2)^2 + (b_1 - b_2)^2]} \leq \eta(\mu)$. In particular:

(34) $|H(\lambda,v,a,b) — H(0,v,a,b)| \leq \mu$; i.e. $|H(\lambda,v,a,b) — h(v,a,b)| \leq \mu$,

for all $<\lambda,v,a,b> \in [-k,k] \times \Delta$ such that $|\lambda| \leq \eta(\mu)$ (see (32) above.) Taking $\mu = \mu_0$ and $\lambda = \lambda_0$, we conclude that:

(35) for all $<v,a,b> \in \Delta$, $|H(\lambda_0,v,a,b) — h(v,a,b)| \leq \mu_0$, provided $|\lambda_0| \leq \eta(\mu_0)$.

Suppose that, as a matter of fact, the numerical value λ_0 satisfies $|\lambda_0| \leq \eta(\mu_0)$. Inequality (35) will then be true for all $<v,a,b> \in \Delta$. In order to explain the empirical success of (28), assume both proposition (29) and $[(Q(v,a,b) \wedge (<v,a,b> \in \Delta)]$. By (29): $H(\lambda_0,v,a,b) = 0$. By (35): $|h(v,a,b)| \leq \mu_0$. Thus (30), which is observationally equivalent to (28), holds good: the truth of the new theory accounts for the confirmation of the old one in Δ.

There is another form of the Correspondence Principle which actually possesses greater heuristic power than (31) (Zahar 1989, Chapter 7, §7.1). It requires that for any $<a,b>$, we have: $|H(\lambda,v,a,b)–h(v,a,b)| \to 0$ as $\lambda v \to 0$. Fixing v and letting $\lambda \to 0$, we obtain: $|H(\lambda,v,a,b)–h(v,a,b)| \to 0$, that is: $H(\lambda,v,a,b) \to h(v,a,b)$, as $\lambda \to 0$. Thus, we retrieve condition (31). But we can alternatively fix λ at $\lambda = \lambda_0$ and let $v \to 0$. This yields: $|H(\lambda_0,v,a,b)–h(v,a,b)| \to 0$ as $v \to 0$. By continuity, we must also have: $|H(\lambda_0,v,a,b)–h(v,a,b)| \to |H(\lambda_0,0,a,b)–h(0,a,b)|$ as $v \to 0$. Hence:

(36) $|H(\lambda_0,0,a,b) — h(0,a,b)| = 0$. That is: $H(\lambda_0,0,a,b) = h(0,a,b)$ for all $<a,b> \in A \times B$.

From (26) we infer:

(37) $(\forall a,b)[Q(0,a,b) \Rightarrow (H(\lambda_0,0,a,b) = h(0,a,b) = 0)]$.

In other words: for v = 0, the new law coincides with the old one.

As an example, we can cite the case of the Special Theory of Relativity, where: $\lambda_0 = 1/c$, c = velocity of light, and v = speed of the mobile. It is assumed that as $(v/c) \rightarrow 0$, the relativistic equation of motion 'tends' to the Newtonian one; where both v and 1/c are now treated as variable parameters. It follows from the above that the classical law $\underline{f}—m\underline{a} = 0$ must strictly hold for v = 0. This desideratum, together with the constraint of Lorentz-covariance, enabled Planck to establish the law $\underline{f} = d[m\underline{v}/\surd(1—v^2/c^2)]/dt$ (see Zahar 1989, Chapter 7).

Let us repeat that the above considerations depend crucially on one topological result; namely on the Heine-Borel theorem which applies to all closed bounded subsets of $\mathbf{R^n}$. This result leaves open the possibility for the two hypotheses, the old one and the new, to diverge widely outside D; for uniform continuity generally holds only over *compact* subsets of $\mathbf{R^n}$.

4.5 The Regulative Role of Symmetry Principles

We have seen that both continuity *and* divergence are of the essence of scientific revolutions; but only continuity has so far been shown to possess any heuristic power. Referring back to the previous section, suppose that Theory (25) has been refuted outside Δ. If no further information is provided, then all we can do is modify (25) in some adhoc fashion. (25) could, for example, be restricted to the domain Δ; that is: instead of (25), (28) could be adopted as our basic theory. But since (28) is logically weaker than (25), a refutation would have been evaded only through an outright *reduction* of the content of (25). This method, which consists in the addition of extra qualifications to the antecedents of existing laws, is known to have been applied by social scientists confronted with certain types of empirical refutation. Physicists also resort to adhoc stratagems, but of a different kind. Starting from an old hypothesis such as (25) above, they construct a law of the form $[H(\lambda,v,a,b) = 0]$; where λ is assumed to be a parameter so small as to make the new theory empirically indiscernible from (25) in the domain Δ. They then adjust λ so as to make it coincide with some value λ_0, where λ_0 is chosen in such a way that $[H(\lambda_0,v,a,b) = 0]$ subsumes the results known to have refuted (25); and the physicists hope that $|\lambda_0|$ will prove to be smaller than the $\eta(\mu_0)$ defined above (4.4).

For example: the later phases of Greek Astronomy offer the example of a programme built on the adhoc method of introducing epicycles specifically doctored to yield the facts which falsified earlier elements of the programme. There was no overall strategy for determining the epicycles independently of known observational results (see Lakatos and Zahar 1978). In such cases, the general feeling among scientists is that the new hypotheses possess no greater truth-likeness than the old ones, which incidentally shows that despite being common to all programmes, the Correspondence Principle falls short of providing a comprehensive heuristic for scientific research. Additional constraints are needed. These often take the form of symmetry conditions imposed on the laws to be constructed. For example: Classical Mechanics, Special Relativity, and General Relativity were required to keep their form respectively under the Galilean transformation, the Lorentz group, and all well-behaved bijections between arbitrary coordinate systems. Another example is provided by Weyl's Unified Field Theory which, beyond the requirement of general covariance, was subjected to a condition of gauge-invariance. Mathematics thus plays a major regulative role by providing symmetry principles under which physical laws must be subsumed.

Symmetries however raise unexpected, not to say paradoxical, problems. No serious difficulties arise in the case of two incompatible theories, which generally have very different symmetry groups; but when an old law is strictly entailed by a new one, we face a problem posed by the so-called Curie-Post Principle (see Redhead 1975, p. 83). The latter asserts that a general proposition transfers its symmetries to its logical consequences and to its special cases. At first sight, this heredity principle seems self-evident: if two situations are not told apart by a strong theory, then they cannot *a fortiori* be distinguished by a weaker one. This has however the unfortunate consequence that as long as it progresses cumulatively, physics can display no new symmetries. For example, since electromagnetism extends electrostatics, the latter should be Lorentz-covariant, which however seems patently false.

The Curie-Post principle implies that no symmetry conditions can help us towards generating new hypotheses, for as long as we try to strengthen existing laws, no symmetry properties *could possibly be* new. Thus it seems as though neither the Correspondence nor any symmetry requirements enable us to move away from existing theories; and the only heuristic method left for strengthening or extending a physical law appears to be that of modifying it—in a purely adhoc way—so as to make it fit recalcitrant facts. Whence the need to demonstrate that the Curie-Post principle can be blocked; for only in this way can the *mathematical conditions* of symmetry and of correspondence be justifiably seen as regulative ideas which might help us towards constructing new hypotheses consistent with the old ones.

Let $\mathbf{K}(\underline{x},a,b,...)$ be a law, where: $\underline{x}$ denotes a sequence of spatio-temporal coordinates, while a, b, and so forth refer to physical quantities such as the components of the electric and magnetic fields. A symmetry of **K** consists of a sequence of functions $<\varphi(\underline{x}), \alpha(\underline{x},a,b,...), \beta(\underline{x},a,b,...),...>$ such that, if we put: $\underline{x}' = \varphi(\underline{x})$, $a' = \alpha(\underline{x},a,b,...)$, $b' = \beta(\underline{x},a,b,...)$, etc, we have:

$$(38)\ \vdash (\forall\underline{x})(\forall a,b,...)[\mathbf{K}(\underline{x},a,b,...) \Leftrightarrow \mathbf{K}(\underline{x}',a',b',...)].$$

For the sake of of simplicity, we shall henceforth write $(\underline{x},a,b)$ for $(\underline{x},a,b,...)$. Thus (38) is shorthand for:

$$(38')\ (\forall\underline{x},a,b)[\mathbf{K}(\underline{x},a,b) \Leftrightarrow \mathbf{K}(\varphi(\underline{x}),\alpha(\underline{x},a,b),\beta(\underline{x},a,b))].$$

Any symmetry can be *passively* construed as a change from one frame Γ to another, Γ' say, such that: $\underline{x}$ and $\underline{x}'$ are the coordinates of the *same* point P, while a and a' denote the values assumed, at P, by the *same* physical entity when referred respectively to Γ and to Γ'. We could alternatively remain within the same frame Γ and *actively* interpret $\underline{x} \rightarrow \underline{x}'$ as a map of Γ into itself which sends a and b into a' and b' respectively.

Note that both construals block Kretschmann's objection that an arbitrary physical law can be rewritten in a generally covariant form; for α and β are defined as quantities dependent *only* on $(\underline{x}, a, b)$. This caveat forbids the introduction of new entities which might trivially make every law form-invariant under *all* well-behaved bijections. Specifying the arguments $\underline{x}$, a, and b of the functions α, β, ... therefore lends heuristic efficacy to the symmetry principles (see Kretschmann 1917).

We are now in a position to prove that a logical consequence $\mathbf{E}(\underline{x},a,b)$ of $\mathbf{K}(\underline{x},a,b)$ need not possess the symmetry $<\varphi,\alpha,\beta>$ of **K**. Suppose that:

$$(39)\ \vdash (\forall\underline{x},a,b)[\mathbf{K}(\underline{x},a,b) \Rightarrow \mathbf{E}(\underline{x},a,b)].$$

Hence:

$$(40)\ \vdash [\mathbf{K}(\underline{x}',a',b') \Rightarrow \mathbf{E}(\underline{x}',a',b')].$$

So, by (38):

$$(40')\ \vdash (\forall\underline{x},a,b)[\mathbf{K}(\underline{x},a,b) \Rightarrow \mathbf{E}(\underline{x}',a',b')].$$

The relations (39) and (40') do not however logically entail $\vdash [\mathbf{E}(\underline{x},a,b) \Leftrightarrow \mathbf{E}(\underline{x}',a',b')]$. As a trivial counter-example, consider the case where **K** is self-contradictory and consequently implies *all hypotheses*. For example, let $\mathbf{K}(\underline{x},a,b) \equiv [(\underline{x} = \underline{x}) \wedge (a = a) \wedge (b \neq b)]$. Since $\mathbf{K}(\underline{x}',a',b') \equiv [(\underline{x}' = \underline{x}') \wedge (a' = a') \wedge (b' \neq b')]$, $\mathbf{K}(\underline{x}',a',b')$ is also inconsistent. Thus (38), that is (38'), holds good for all φ, α, and β, which means that **K** admits all functions $\langle\varphi, \alpha, \beta\rangle$ as symmetries. It is however far from true that any $\langle\varphi, \alpha, \beta\rangle$ defines a symmetry for every physical theory. We can, for instance, point out that Classical Mechanics is not Lorentz-covariant.

Assuming the truth of (38), it is just as important for us to prove that the 'special cases' of the hypothesis $\mathbf{K}(\underline{x},a,b)$ need not be symmetric under $\langle\varphi,\alpha,\beta\rangle$. To this end, it proves convenient to regard every symmetry as a map of the set Ω of all the solutions of **K** into itself. By definition, Ω is the class of all tuples $\langle A(\underline{x}), B(\underline{x})\rangle$ of functions of $\underline{x}$ such that:

(41) $\vdash (\forall \underline{x})\mathbf{K}(\underline{x}, A(\underline{x}), B(\underline{x}))$. By (38'):
(42) $\vdash (\forall \underline{x})\mathbf{K}(\varphi(\underline{x}), \alpha(\underline{x}, A(\underline{x}), B(\underline{x})), \beta(\underline{x}, A(\underline{x}), B(\underline{x})))$.

Substituting $\varphi^{-1}(\underline{x})$ for $\underline{x}$ and then generalizing, we obtain:

(43) $\vdash (\forall \underline{x})\mathbf{K}(\underline{x}, A'(\underline{x}), B'(\underline{x}))$, where: $A'(\underline{x}) =_{Def.} \alpha(\varphi^{-1}(\underline{x}), A(\varphi^{-1}(\underline{x})), B(\varphi^{-1}(\underline{x})))$ and:

$$B'(\underline{x}) =_{Def.} \beta(\varphi^{-1}(\underline{x}),A(\varphi^{-1}(\underline{x})), B(\varphi^{-1}(\underline{x}))).$$

Taken together, (41) and (43) tell us that if $\langle A(\underline{x}), B(\underline{x})\rangle$ is a solution of the law **K**, then due to the symmetry $\langle\varphi,\alpha,\beta\rangle$ of **K**, $\langle A'(\underline{x}), B'(\underline{x})\rangle$ will constitute another solution of the same law. We can in fact interpret this symmetry as the application of Ω into itself defined by:

$\langle A(\underline{x}), B(\underline{x})\rangle \rightarrow \langle A'(\underline{x}), B'(\underline{x})\rangle$, where: $\langle A'(\underline{x}), B'(\underline{x})\rangle =_{Def.}$ $\langle\alpha(\varphi^{-1}(\underline{x}), A\varphi^{-1}(\underline{x}), B\varphi^{-1}(\underline{x})), \beta(\varphi^{-1}(\underline{x}), A\varphi^{-1}(\underline{x}), B\varphi^{-1}(\underline{x}))\rangle$.

It is easily seen that this map does not necessarily send a 'special case', that is, a solution of a given type, into another one of the same type. For example: let Ω_0 be the subset of Ω consisting of all tuples of the form $\langle f(\underline{x}), g(\underline{x})\rangle$, where $g(\underline{x}) = 0$ = zero function. Thus:

(44) $\vdash (\forall \underline{x})\mathbf{K}(\underline{x}, f(\underline{x}), 0)$.

The symmetry $\langle\varphi,\alpha,\beta\rangle$ sends $\langle f(\underline{x}), g(\underline{x})\rangle$ into $\langle f'(\underline{x}), g'(\underline{x})\rangle$, where: $g'(\underline{x}) = \beta(\varphi^{-1}(\underline{x}), f(\varphi^{-1}(\underline{x})), g(\varphi^{-1}(\underline{x}))) = \beta(\varphi^{-1}(\underline{x}), f(\varphi^{-1}(\underline{x})), 0)$ since g is the zero function. There is however no reason to suppose that g' vanishes identically; so that we might well have $\langle f'(\underline{x}), g'(\underline{x})\rangle \notin \Omega_0$. Here is an example given by Michael Redhead in his seminal paper on the role of symmetry in physics (Redhead 1975, pp. 103–04). Apply Maxwell's equations to a static system of charged particles. Taking $f(\underline{x})$ and $g(\underline{x})$ to be the electric and the magnetic fields respectively, we conclude that $g(\underline{x}) = 0$. Let $\varphi(\underline{x})$ be any non-trivial Lorentz transformation. We know that Maxwell's theory is Lorentz-covariant and hence form-invariant under the symmetry φ. As already explained, φ can be *passively* viewed as a change from a stationary frame to a mobile one, in which the particles are no longer at rest. The moving frame will therefore contain both an electric and a non-vanishing magnetic field; that is: $g'(\underline{x}) \neq 0$. The heredity principle is consequently false. This entitles us to search for hitherto unknown symmetries to be imposed on new hypotheses; where the latter *might*, for all we know, turn out to be consistent with our past conjectures. But how are we to find such symmetries? Going back to (38'), it is obvious that if **K,** φ, α, and β were all unknown functions, then it would be impossible for us even to begin our search. We could however start by considering a *given* theory **K** in order then to determine the class of its symmetries $<\varphi, \alpha, \beta>$; or from some *available* $<\varphi, \alpha, \beta>$, which could then be heuristically exploited in the construction of an appropriate **K**, that is: of a theory covariant under $<\varphi, \alpha, \beta>$. And note that even within the same programme, these two approaches need not be mutually exclusive. For example: **K** could form part of a more general theory; it could for instance be the set of Maxwell's equations taken within the context of classical physics. After examining the symmetries of **K**, we might find that they do not subsume the whole system in which **K** is imbedded. Thus, Classical Mechanics is not Lorentz-covariant since it does not possess all the Maxwellian symmetries. The latter can be put to heuristic use by becoming adequacy requirements to be imposed on *all* the laws of nature. Both Einstein and Poincaré started from a known **K**, namely from Maxwell's equations. Both singled out the symmetry group *L* of **K**, and both used *L* in order to generate new hypotheses. This is also how Einstein and Planck modified the old laws of motion into Lorentz-covariant relations. As for Poincaré, apart from constructing a Lorentz-covariant dynamics, he used the Principles of Correspondence and of Lorentz-invariance in order to build a new relativistic theory of gravitation (Poincaré 1906b; and see below, 4.8).

Another heuristic device consists in strengthening a covariance requirement by enlarging an existing symmetry group. For example, both for scientific and for philosophical reasons, Einstein decided to construct a generally covariant theory of gravitation. His field equations had therefore to keep their form, not only under the Lorentz transformations, but also under all well-behaved bijections (Einstein 1923). Such a move seems incidentally to be forbidden by the Curie-Post principle, which is one reason why the latter had to be confined within strict limits.

The question finally arises as to the existence of any connection between the heuristic importance of mathematics on the one hand, and metaphysical realism on the other. To my mind, the cautious answer should be: yes, there is such a connection, but only an indirect one based exclusively on plausibility considerations. It would be very surprising if the mathematical form of **K**, which plays a crucial heuristic role and is imbedded—if only approximately—in all subsequent developments of physics, should reveal nothing about the structure of reality. The quasi-cumulative evolution of these mathematical forms, which moreover accounts for the growth of technological knowledge, speaks against any such sceptical conclusion. To think otherwise would, in Poincaré's own words, attribute too great an importance to blind chance. Taken together with the ineliminability of mathematics from physical systems, this plausibility argument provides considerable support for scientific realism, at least in its structural version (see above, Chapter 2).

4.6 The Reform of Electromagnetism

Let us give a synopsis of the conclusions reached in 4.3 above. Together with Einstein but independently of him, Poincaré launched the Relativity Programme. On the one hand, his empiricist criterion of meaning led him towards an operational definition of Lorentz's local time t'' (see above, Chapter 1). t'' is the time measured by clocks which are synchronised according to the Poincaré-Einstein convention: let a light signal leave some point A at A-time t_1 and arrive, at B-time t_2, at another point B where it is instantaneously reflected before going back and reaching A at A-time t_3. The two clocks at A and B are said to be synchronous if and only if $t_2 = (t_1 + t_3)/2$. On the other hand, Poincaré rid physics of the Galilean Transformation which Lorentz had always interposed between the ether frame and the effective coordinates—between (x, y, z, t) and (x', y', z', t'), where: $x' = \gamma(x - ut)$, $y' = y$, $z' = z$, $t' = \gamma(t - ux/c^2)$ and $\gamma = 1/\sqrt{(1 - u^2/c^2)}$.

In 'Sur la Dynamique de l'Électron', published in 1905–1906, Poincaré looks upon the effective coordinates as the only physically significant parameters. In opposition to Lorentz, he was not wedded to the classical ontology according to which absolute time, the ether frame and the Galilean coordinates are the only intelligible entities. This is because, in accordance with Poincaré's conventionalism and with his empiricist meaning criterion, hypotheses are neither true nor false but more or less convenient. The degree of convenience, that is: of the mathematical perfection of a system can nevertheless be treated as an *index of its verisimilitude.* Through its global structure, a convenient theory can simulate an objective reality which it cannot directly denote. Despite their metaphorical character, hypotheses can therefore be more or less verisimilar.

As already mentioned, Poincaré was a Kantian of sorts: scientific hypotheses can mirror the noumena without being susceptible of any direct reference to them (see above, Chapters 1 and 2). From a methodological viewpoint, both absolute time and the ether hypothesis should be regarded as temporary devices whose function is to assist our imagination; since they do not reveal the deep nature of things, they ought to be rejected as soon as they prove cumbersome. Because the Galilean transformation had become superfluous, Poincaré brushed it aside in favour of the Lorentz-Transformation which takes us directly from the rest-frame to the effective variables. Let us also recall that he formulated the heuristic on which STR was to be based: every physical law is subject to two conditions; on the one hand, it ought to be covariant under the Lorentz group—it must assume the same form in all inertial frames—and on the other, it should, wherever possible, tend towards some corresponding classical law when the envisaged velocities become 'small' in comparison with c (see 4.4 above). Thus the two pillars of the STR programme are: the Correspondence Principle and a symmetry requirement given in the form of a Lorentz-covariance condition (see 4.5 above).

Poincaré used the new heuristic in order to reconstruct and also to correct Lorentz's transformation rules for the electromagnetic field (see equations (7) and (9) above). He began, in effect, by considering the following set of relations:

(23) $c' = c/(\Omega\psi\gamma)$ and

(24) $x' = \gamma(x-ut)/\psi$; $y' = y/\psi$; $z' = z/\psi$; $t' = \Omega\gamma^2(t - ux/c^2)$.

Let us now put:

(45) $x^* = \psi x'$; $y^* = \psi y'$; $z^* = \psi z'$ and $t^* = t'/(\Omega\gamma)$.

Thus:

(46) $x^* = \gamma(x - ut)$; $y^* = y$; $z^* = z$; $t^* = \gamma(t - ux/c^2)$.

In other words, (x^*, y^*, z^*, t^*) are obtained from (x, y, z, t) by means of the familiar Lorentz-Transformation (LT). Substituting $-u$ for u, then interchanging the starred with the unstarred coordinates, we determine the inverse transformation:

(47) $x = \gamma(x^* + ut^*)$; $y = y^*$; $z = z^*$; $t = \gamma(t^* + ux^*/c^2)$.

In terms of (x', y', z', t'):

(48) $x = \gamma[\psi x' + ut'/(\Omega\gamma)]$; $y = \psi y'$; $z = \psi z'$; $t = \gamma[(t'/(\Omega\gamma)) + \psi ux'/c^2]$.

Rather than follow Poincaré's own methods which are based on the structure of the electromagnetic 4-vector potential, we shall derive the transformation of the field directly from equations (24). This approach is justified by the fact that Poincaré in effect used (24) as the starting point of his investigations. Write: $\nabla^* = (\partial/\partial x^*, \partial/\partial y^*, \partial/\partial z^*)$. By (45):

(49) $\nabla^* = (1/\psi) \cdot (\partial/\partial x', \partial/\partial y', \partial/\partial z') = (1/\psi) \cdot \nabla'$. Moreover: $\partial/\partial t^* = \Omega\gamma \cdot \partial/\partial t'$.

Now define the two quantities $\underline{E}^*$ and $\underline{H}^*$ as follows:

(50) $\underline{E}^* = (E_1, \gamma(E_2 - uH_3/c), \gamma(E_3 + uH_2/c))$; $\underline{H}^* = (H_1, \gamma(H_2 + uE_3/c), \gamma(H_3 - uE_2/c))$.

These are equations (9) which were discovered by Lorentz and with which Poincaré was no doubt familiar. All we need to do is determine a transform ρ^* of the charge density ρ in such a way that Maxwell's theory keeps its form under (46) and (49). Remember that:

(1) $\nabla \cdot \underline{E} = \rho$; $\nabla \cdot \underline{H} = 0$; $\nabla \times \underline{E} = -[\partial\underline{H}/\partial t]/c$; $\nabla \times \underline{H} = [(\partial\underline{E}/\partial t) + \rho\underline{v}]/c$;

where we put:

(51) $\underline{v} = (v_1, v_2, v_3)$ [= (dx/dt, dy/dt, dz/dt)] = charge velocity.

Now note that by (45): $(dx^*/dt^*, dy^*/dt^*, dz^*/dt^*) =$

$\Omega\psi\gamma(dx'/dt', dy'/dt', dz'/dt')$. Thus:

(52) $\underline{v}^* = (v_1^*, v_2^*, v_3^*) = \Omega\psi\gamma(v_1', v_2', v_3') = \Omega\psi\gamma \cdot \underline{v}'$;

where $\underline{v}^*$ and $\underline{v}'$ are the 'charge' velocities corresponding to $\underline{v}$. Putting:

(53) $\rho^* = \rho\gamma(1 - uv_1/c^2)$,

we can verify, in the usual manner, that:

(54) $\nabla^* \cdot \underline{E}^* = \rho^*$; $\nabla^* \cdot \underline{H}^* = 0$; $\nabla^* \times \underline{E}^* = -(1/c)\partial\underline{H}^*/\partial t^*$; $\nabla^* \times \underline{H}^* = (1/c)[(\partial\underline{E}^*/\partial t^*) + \rho^*\underline{v}^*]$.

It follows from (49) and (53) that:

(55) $\nabla' \cdot \underline{E}^* = \psi\rho^* = \psi\rho\gamma(1-uv_1/c^2)'$; $\nabla'. \cdot \underline{H}' = 0$; $\nabla' \times \underline{E}^* = -(\Omega\psi\gamma/c)\partial\underline{H}^*/\partial t' = -(1/c')\partial\underline{H}^*/\partial t'$ [by (23)]; $\nabla' \times \underline{H}^* = (\psi/c)[(\Omega\gamma\partial\underline{E}^*/\partial t') + \rho^*\psi\Omega\gamma\underline{v}'] = (1/c')[(\partial\underline{E}^*/\partial t') + \rho^*\psi\underline{v}']$.

Take k to be a constant, which we temporarily leave unspecified, and put:

(56) $\underline{E}' = k\underline{E}^* = k(E_1, \gamma(E_2 - uH_3/c), \gamma(E_3 + uH_2/c))$; $\rho' = k\psi\rho^* = k\psi\rho\gamma(1 - uv_1/c^2)$; $\underline{H}' = k\underline{H}^* = (H_1, \gamma(H_2, + uE_3/c), \gamma(H_3 - uE_2/c))$.

It follows from the above that:

(57) $\nabla' \cdot \underline{E}' = \rho'$; $\nabla' \cdot \underline{H}' = 0$; $\nabla' \times \underline{E}' = -(1/c')\partial\underline{H}'/\partial t'$; $\nabla'\times\underline{H}' = (1/c')[(\partial\underline{E}'/\partial t') + \rho'\underline{v}']$.

This means that relative to the variables (x', y', z', t'), Maxwell's equations keep their form. We should however note that although the velocity c' of wave-propagation is a constant, it might differ from c; that is: we may well have $c \neq c/(\Omega\psi\gamma) = c'$; so that despite being a constant throughout the 'moving' frame, c' is not necessarily an invariant.

k has so far been treated as an arbitrary constant. Using a method introduced by Poincaré, we shall prove that k is fixed by the the invariance law of the electric charge. In the 'stationary' frame, consider a charge e enclosed by a mobile sphere S_0 which has radius r and moves with uniform velocity (v_1, v_2, v_3). The charge-density (is therefore given by:

(58) $\rho = 3e/(4\pi r^3)$, while the equation of S_0 has the form:

(59) $[x - (v_1t + b_1)]^2 + [y - (v_2t + b_2)]^2 + [z - (v_3t + b_3)]^2 = r^2$.

Substituting for (x, y, z, t) in terms of (x', y', z', t'), we obtain the equation of S_0 in the 'moving' frame which, by virtue of (48), turns out to be a relation of the form:

(60) $[\gamma\psi(1 - uv_1/c^2)x' + h_1t' + d_1]^2 + [\psi y' + s_2x' + h_2t' + d_2]^2 + [\psi z' + s_3x' + h_3t' + d_3]^2 = r^2$.

This is the equation of an ellipsoid whose volume is equal to $4\pi r^3/[3\gamma\psi(1 - uv_1/c^2)\psi\psi] = 4\pi r^3/[3\gamma\psi^3(1 - uv_1/c^2)]$. Assuming the charge e to be the same in the moving as in the stationary frame, we conclude that:

(61) $\rho' = 3\gamma\psi(^3(1 - uv_1/c^2)e/(4\pi r^3) = \rho\gamma\psi^3(1 - uv_1/c^2)$, by (58).

Comparing with the expression of ρ' given by (56) above, we infer:

(62) $k = \psi^2$. Our transformation equations (24) and (56) can finally be written as:

(63) $x' = \gamma(x - ut)/\psi$; $y' = y/\psi$; $z' = z/\psi$; $t' = \Omega\gamma^2(1 - ux/c^2)$;

where: $\gamma = 1/\sqrt{(1 - u^2/c^2)}$; $\underline{E}' = \psi^2(E_1, \gamma(E_2 - uH_3/c), \gamma(E_3 + uH_2/c))$; $\underline{H}' = \psi^2(H_1, \gamma(H_2 + uE_3/c), \gamma(H_3 - uE_2/c))$; $\rho' = \psi^3\gamma(1 - uv_1/c^2) \cdot \rho$; and $c' = c/(\Omega\psi\gamma)$ = speed of light in moving frame.

Note that the ordinary LT is obtained by taking $\psi = \Omega\gamma = 1$ in (63).

4.7 Poincaré's 4-Dimensional World

Well before Minkowski, Poincaré conceived of a physical frame as an indivisible 4-dimensional continuum: one in which the time variable t is transformed *alongside* the three spatial coordinates x, y, and z (see (24) above).

Going back to (23) and (24), we shall henceforth assume that c is an invariant, in other words that $c = c' = c/(\Omega\psi\gamma)$. Thus $\Omega = 1/(\psi\gamma)$. We moreover choose our units in such a way that c = 1. Putting: $1/\psi = h$ and $\beta = -u/c = -u$, we have: $\Omega = 1/(\psi\gamma) = h/\gamma$ where $\gamma = 1/\sqrt{(1 - \beta^2)}$. Therefore, by (63):

(64) $x' = h\gamma(x + \beta t)$; $y' = hy$; $z' = hz$; $t' = h\gamma(t+\beta x)$.

It can be easily verified that:

(65) $h^2(x^2 + y^2 + z^2 - t^2) = (x'^2 + y'^2 + z'^2 - t'^2)$. Hence 1, that is, c, is an invariant speed.

It is worth noting that the Newtonian world is also 4-dimensional since events occurring in it have both spatial and temporal coordinates. But in view of (5) $x_g = x - ut$; $y_g = y$; $z_g = z$; $t_g = t$, the time t is detached from the spatial coordinates (x, y, z) in such a way that δt is the basic invariant. In Newton's mechanical science as distinct from his metaphysics, time is so to speak more absolute than space. It makes *scientific* sense to say of two events that they are simultaneous, but not that they occur at the same place; for by virtue of (5), we might have: $u \neq 0$, $x_1 = x_2$ and $t_1 \neq t_2$; so that $x_{g1} \neq x_{g2}$. In other words, if two spatially separated events are simultaneous in one system of coordinates, then they are so in all Galilean frames; two non-simultaneous events may however happen at the same spatial point in one such frame but not in all others.

Let us now go back to Transformation (64), which will henceforth be denoted by T(β, h). β and h are provisionally treated as independent parameters. Define:

(66) $H = \{\ T(\beta,h): \beta \in \mathbf{R},\ 0 \neq h \in \mathbf{R}\ \}$.

It can be verified that H forms a group. T(0,1) trivially turns out to be the identity map, which must therefore lie in H. Solving (64) for (x, y, z, t) in terms of (x', y', z', t'), we obtain:

(67) $x = \gamma(x' - \beta t')/h;\ y = y'/h;\ z = z'/h;\ t = \gamma(t' - \beta x')/h$.

That is:

(68) $[T(\beta,h)]^{-1} = T(-\beta,1/h) \in H$. Finally, it can be easily, though laboriously, verified that:

(69) $T(\beta',h').T(\beta,h) = T((\beta' + \beta)/(1 + \beta'\beta),\ h'h) \in H$;

where the product on the left-hand side denotes the composition operation of two functions, i.e. as the operation of carrying out, first T(β,h), and then T(β',h'). Being closed under the operations of composition and of taking inverses, H is a group of transformations.

We shall henceforth write $\underline{r}$ and $\underline{r}_i$ for (x, y, z) and (x_i, y_i, z_i), also $\underline{r}^2$ and $\underline{r}_i^2$ for $x^2 + y^2 + z^2$ and $x_i^2 + y_i^2 + z_i^2$ respectively, where i is any subscript (or superscript). Poincaré defines the Lorentz group as the set L of all non-singular homogeneous linear transformations which multiply $\underline{r}^2 - t^2$ by a positive constant. Let us show that L is actually a group. Consider two functions f_1 and f_2 belonging to L; put: $f_1(\underline{r},t) = (\underline{r}_1,t_1)$ and $f_2(\underline{r}_1,t_1) = (\underline{r}_2,t_2)$. Thus $f_1^{-1}(\underline{r}_1,t_1) = (\underline{r},t)$. By definition of L, there exist two constants h_1 and h_2 such that: $\underline{r}_1^2 - t_1^2 = h_1^2(\underline{r}^2 - t^2)$ and $\underline{r}_2^2 - t_2^2 = h_2^2(\underline{r}_1^2 - t_1^2)$; thus $\underline{r}^2 - t^2 = (h_1^{-1})^2(\underline{r}_1^2 - t_1^2)$, so $f_1^{-1} \in L$; moreover: $f_2f_1(\underline{r},t) = f_2(\underline{r}_1,t_1) = (\underline{r}_2,t_2)$, where: $\underline{r}_2^2 - t_2^2 = h_2^2(\underline{r}_1^2 - t_1^2) = (h_2h_1)^2(\underline{r}^2 - t^2)$, so that $f_2f_2 \in L$. Being closed under the operations of function composition and of taking inverses, L constitutes a group. (65) moreover implies that H is a subset and hence a subgroup of L, a result which we express by

(70) $H \leq L$.

So far, we have looked upon β and h as two independent parameters. It is now time to take account of the physical meaning of the T(β,h)'s, that is: of the transformations which effect the transition from one inertial system to the next. Remember that $\beta = -u/c = -u$, where u is the velocity of the frame [x',y',z',t'] with respect to [x,y,z,t]. This leads to the conjecture that the quantity h occurring in (64) depends on the speed $-\beta$ of the mobile frame. More precisely, let $f(\beta)$ be a non-vanishing function of β, i.e. suppose that $f(\beta) \neq 0$ for all $\beta \in \mathbf{R}$. Taking $h = f(\beta)$ in (64), we obtain:

(71) $x' = f(\beta) \cdot \gamma \cdot (x + \beta t);\ y' = f(\beta) \cdot y;\ z' = f(\beta) \cdot z;\ t' = f(\beta) \cdot \gamma \cdot (t + \beta x).$

Equations (71) represent the transformation $T(\beta, f(\beta))$ which connects the inertial frame [x,y,z,t] with the system [x',y',z',t'] moving with uniform velocity $-\beta$ along the x-axis of [x,y,z,t]. Now define:

(72) $K = \{\ T(\beta, f(\beta)) \colon \beta \in \mathbf{R}\ \} \subseteq H \leq L.$

Thus the members of K permute the inertial frames any two of which are declared by the Relativity Postulate to be physically equivalent to each other. K must therefore be a group. From this result, let us infer that for all $\beta \in \mathbf{R}$, $f(\beta) = 1$. By definition of K, $T(\beta, f(\beta)) \in K$; so we must also have: $[T(v, f(\beta))]^{-1} \in K$. Thus, by (68), $T(-\beta, 1/f(\beta)) = [T(\beta, f(\beta))]^{-1} \in K$. By definition of K, it follows that: $f(-\beta) = 1/f(\beta)$. Hence:

(73) $f(\beta).f(-\beta) = 1$, for all $\beta \in K$.

Now go back to (71) above. Let $[\xi, \eta, \zeta, ($] and $[\xi', \eta', \zeta', \tau']$ be the coordinate systems obtained by rotating the frames [x,y,z,t] and [x',y',z',t'] through an angle π about the y- and y'-axes respectively. Thus:

(74) $\xi = -x;\ \eta = y;\ \zeta = -z;\ \tau = t;\ \xi' = -x';\ \eta' = y';\ \zeta' = -z';\ \tau' = t'.$

Hence, by (71):

(75) $\xi' = -x' = -f(\beta) \cdot \gamma[x + \beta t] = f(\beta) \cdot \gamma[-x - \beta t] = f(\beta) \cdot \gamma[\xi - \beta\tau].$

The equation $\xi' = 0$, that is $\xi - \beta\tau = 0$, gives the velocity of the frame $[\xi', \eta', \zeta', \tau']$ relative to $[\xi, \eta, \zeta, \tau]$, which is thus equal to β. By (71), it follows that:

(76) $\xi' = f(-\beta) \cdot \gamma[\xi - \beta\tau];\ \eta' = f(-\beta) \cdot \eta;\ \zeta' = f(-\beta) \cdot \zeta;\ \tau' = f(-\beta) \cdot \gamma[\tau - \beta\xi].$

Comparing the last member of (75) with the first equation in (76), we conclude: $f(\beta) = f(-\beta)$. Hence, by (72):

(77) $[f(\beta)]^2 = 1.$

Poincaré chooses the value:

(78) $f(\beta) = +1.$

It follows that: $K \leq H \leq L$. (Note that $f(\beta) = -1$ would contradict the closure of the group K with regard to multiplication; for, by (69):

$T(\beta_1, -1) \cdot T(\beta_2, -1) = T((\beta_1 + \beta_2)/(1 + \beta_1\beta_2),1)$,

where the right-hand side is not of the form $T(\beta_0, -1)$.)

4.8 A Synopsis of Poincaré's Gravitational Theory

Poincaré based his approach to gravitation on an analogy with electromagnetism. Let us recall that since the speed of light is taken to be equal to 1, the Lorentz-force $\underline{L}$ which acts on a particle carrying charge e and moving with velocity $\underline{v}$ is given by:

(79) $\underline{L} = e[\underline{E} + \underline{v} \times \underline{H}]$.

Now consider the equation:

(80) $\underline{E} + \underline{v}\times\underline{H} = d[m\underline{v}/\sqrt{(1 - v^2)}]/dt$.

Poincaré started by proving that relation (80) is Lorentz-covariant—that it keeps its form under all transformations belonging to L. By virtue of an analogy with $\underline{L}$, he then constructed a relativistic theory of gravitation which was determined to within certain coefficients; he finally selected, as values of these coefficients, the simplest Lorentz-invariants which would make the new equations tend to Newton's inverse-square law for small velocities of the test particles. In this final section of the present chapter, I shall content myself with describing the main aspects of Poincaré's gravitational hypothesis; the reader interested in a more rigorous and comprehensive approach is referred to my 1989 (Chapter 5, Section 11.3). Since gravitation treats of the mutual attractions of material bodies, Poincaré set out to determine the simplest invariants which can be built from the velocities and relative positions of any two point-masses. Let P and P_1 be two unit mass-particles and $\underline{G}$ the gravitational force exerted by P_1 on P. $(\xi,\eta,\zeta,\tau,)$ and $(\xi + x,\eta + y,\zeta + z,\tau + t)$ will be the coordinates of P and P_1 respectively. Since physical disturbances are to be propagated with finite speed, $\tau + t$ must precede τ; that is: $t < 0$. Put $\underline{r} = (x,y,x)$ and $r = \sqrt{(x^2 + y^2 + z^2)}$. It comes as no surprise that the velocity of light, 1, is also that of all gravitational actions. Hence $t = -r/c = -r$. Denoting by $\underline{v}$ and $\underline{v}_1$ the velocities of P and P_1 at the instants τ and τ+t respectively, Poincaré found the expression of $\underline{G}$ to be of the form:

(81) $\underline{G} = \lambda[\underline{r} + r\underline{v}_1 + \underline{v} \times (\underline{v}_1 \times \underline{r})]$;

where λ must be chosen in such a way that the new theory tends towards Newton's inverse-square law for small values of v and v_1. The two simplest candidates turn out to be:

(82) $\lambda = -\sqrt{(1 - v^2)} \cdot \sqrt{(1 - v_1^2)^3}/[(r + \underline{r} \cdot \underline{v}_1)^3 \cdot (1 - \underline{v} \cdot \underline{v}_1)]$, and:

(83) $\lambda = -(1 - v_1^2)/(r + \underline{r} \cdot \underline{v}_1)^3$.

In his 1906b, Poincaré does not dwell on the question of a possible dependence of the gravitational mass of a particle on its speed. He is nonetheless aware of the importance of the equivalence principle, according to which the inertial and the gravitational masses of any material body are equal. This principle entails that irrespective of their physical composition, all test-particles experience the same acceleration when placed in the same gravitational field. Unlike Einstein, Poincaré does not however attribute absolute validity to the equivalence of gravity and inertia; he only requires that this principle be a limiting case of some consequence of his new hypothesis. This may well be the reason why he could keep STR as a suitable framework for his gravitational theory. By the same token, it was Einstein's total adherence to the equivalence principle which eventually led him to abandon flat space and adopt Riemannian geometry as a basis of his revolutionary theory of gravitation. In conclusion: Poincaré used his two-pronged heuristic, namely correspondence and symmetry, in order to correct and to reconstruct Lorentz's transformation rules for the electromagnetic field; he thereby developed a new kinematics which, together with the conservation law for the electric charge, enabled him to establish the full Lorentz-covariance of Maxwell's equations. By 1905, he had moreover gone well beyond Lorentz's and even Einstein's achievements; for he had already proposed a theory of gravitation which predicts the propagation of gravitational waves at the speed of light. He finally asked a question which also occurred to Einstein; namely: why should the velocity of light be *the basic invariant* and thus play a fundamental role not only in electrodynamics, but also in all the other branches of physics? To this question he imagined two possible answers: either all forces ultimately have some common electromagnetic origin; or else, it is the observer himself who inserts the velocity of light into Nature by using light signals in order to measure both the temporal and the spatial intervals. Poincaré thus envisaged two answers of very different philosophical character; the first is realistic, and the second at least partly conventionalistic. Until the very end, he remained aware of a tension between the scientist who creates conventions in order to interpret nature, and Nature which sets definite limits to the otherwise free activity of the scientist.

5

The Foundations of Mathematics According to Poincaré and Russell

5.1 An Overview of Poincaré's Philosophy of Mathematics

In his preface to Dugac's book on Dedekind's work, Dieudonné refers to the profound intuition which guided the creative work of great mathematicians like Poincaré, Lebesgue, and Dedekind. Dieudonné claims that their sound mathematical instinct "led them to be wary of the excessive freedom demanded by this [Cantorian] school, a freedom which appeared to be in danger of degenerating into licence" (Dugac 1976, Preface; my translation). In the present chapter, I propose to examine another of Poincaré's powerful instincts, namely his *philosophical* intuition with regard to the foundations of mathematics. We should note that his philosophy of mathematics has been largely—and unjustly—ignored mainly because it conflicts with his actual mathematical *practice*. In previous chapters, we noted that Poincaré not only failed to unify his philosophical views into a coherent system; but also that he frequently shifted his position, thus exhibiting the kind of 'opportunism' which was regarded by Einstein as indispensable to all great scientists. Putting it pedantically, the set of all opinions expressed by Poincaré at different points of his career is inconsistent. Let us however remember that this remark applies to practically all great philosophers, which explains why system-building on a grand scale has gone out of fashion; for such systems are likely at any moment to succumb to their inner contradictions. A good philosopher is nowadays taken to be someone who, by means of analyses guided by a central idea, illuminates or solves some important foundational problem. In this sense, Poincaré was not only a supreme mathematician but also a major philosopher; so that Popper was largely justified in calling him the greatest philosopher of science ever (see Bouveresse 1998, p. 191).

In his controversy with Couturat, Poincaré adopted the haughty and even sarcastic style of a great mathematician irritated by a philosopher's intrusion into areas which ought to be none of an amateur's business. In his articles on the logical paradoxes where he defended his viewpoint as well as those of Peano and Couturat, Russell—for his part—proved capable of the kind of English irony which was in no way inferior to Poincaré's.

We shall now examine a long exchange in which a total reversal of roles took place, namely the debate which opposed Poincaré, on the one hand, to Russell and the logicists on the other. In this situation, it was Poincaré's turn to behave like a 'mere' philosopher vis-à-vis professionals who had created rigorous new disciplines, namely: mathematical logic, the theory of types, and axiomatic set theory. Precisely because by 'logic' Poincaré invariably meant the Aristotelian syllogistics which had been superseded by the new logic, and also because he never bothered to master this modern symbolism and even took a perverse pride in not speaking 'Peanian', some authors have concluded that foundational problems were the chink in Poincaré's scientific armour. Ironically, these authors judged Poincaré by the same standards which he invoked in dismissing Couturat. Despite issuing in some justified criticisms of Poincaré's views, this attitude seems to me fundamentally mistaken. From a methodological standpoint, Russell may have been justified in opposing some of Poincaré's theses; but the high degree of technicality displayed by Russell's system masked flaws which only detailed *philosophical analyses* could expose. Poincaré's philosophical position actually turns out to be very similar to Russell's; but the highly complex framework of ramified type theory conceals weaknesses which only Poincaré could identify in an intuitive manner. Bernays and Gödel were later to rediscover criticisms which Poincaré had levelled at both Zermelo and Russell. If, after having somewhat forgotten Poincaré, we are nowadays more inclined to listen to Gödel, it is for the wrong-headed 'reason' that Gödel uses a logical machinery which Poincaré refused to acquire.

Let me try to characterize, in very general terms, Poincaré's position with regard to the foundations of mathematics. He was, to my mind, neither a pre-intuitionist nor a strict constructivist but a quasi-Kantian. More precisely: he was a constructivist only in the Kantian sense of demanding that mathematicians construct their concepts in the pure intuition of time. But also like Kant, Poincaré accepted the necessity of synthetic a priori principles which are of a level so high as to violate all versions of constructivism. He was for example prepared to accept Russell's Axiom of Reducibility which had driven its author to logical despair. We should moreover note that Poincaré was not a *neo*-Kantian in the Marburg sense of the word. His *quasi*-Kantianism arose from his conviction that Kant's philosophy should not be abandoned but modified in the light of such scientific breakthroughs as the discovery of non-Euclidean geometries, of their equi-consistency with the Euclidean system and of their applicability to physics. Instead of holding the Kantian categories to be constitutive of experience, he adopted a phenomenalist attitude with respect to all 'crude facts', that is: with respect to the data given directly in sense-experience. He furthermore rejected the alleged a priori character of geometry, but like Kant and the later intuitionists, he regarded the subjective intuition of time as the foundation of arithmetic.

As is well known, Kant took the thing-in-itself to be a totally unknowable *existent*; so that we can pretend only to a knowledge of appearances some of whose components, being contributed by the understanding, are infallible. Poincaré similarly held that in the physical sciences we can adopt conventions which can be immunized against empirical refutation. To the extent that these conventions prove systematically convenient, they will however *reflect* an ontological order which they cannot directly signify: they establish true relations between entities whose essence cannot be fathomed. As for mathematical knowledge, one often has the impression that Poincaré shared Hermite's views which he described as follows:

> I have not known a more realistic mathematician, in the Platonic sense, than Hermite, and yet I must admit that I have never met one more opposed to Cantorism. There is in this a seeming contradiction, all the more so since he repeated readily: "I am anti-Cantorian because I am a realist." He blamed Cantor for creating objects instead of being content to discover them. It was no doubt due to his religious convictions that he considered it a sort of impiety to wish to penetrate without difficulty a domain which God alone can comprehend and not to wait until He reveals its mysteries to us one by one. He compared the mathematical sciences to the natural sciences. A naturalist who sought to guess God's secret instead of learning through experience would have seemed to him not only presumptuous but disrespectful of the Divine Majesty. To him the Cantorians seemed to want to act in the same manner in mathematics. And that is why he was an idealist in practice whereas he was a realist in theory. There is one reality to be known and it is exterior to us and independent of us; but all we can know of it depends on us, and is then merely a becoming, a kind of stratification of successive conquests. The rest is real but eternally unknowable. (Poincaré 1913b, p. 73)

This passage shows that Poincaré's constructivism had an epistemological rather than a metaphysical character. As a Kantian, he might in the end have accepted the existence of such mind-independent 'noumena' as the natural and even the real numbers. He would nonetheless have insisted that we have no direct access to these things-in-themselves, which is why we should always try to give constructive and hence predicative proofs of all mathematical theorems. A constructed entity exists at least in the sense of being captured by some operations of the human mind. Since human constructions can certainly be carried out by God, Hermite's argument shows that atemporal Platonic Forms *might* correspond to these constructs; and these Forms may include such non-constructible entities as the transfinite cardinals. But of these we shall know nothing except in some roundabout way; for instance through their being incorporated in theories which provide new solutions to concrete problems. Note that for Poincaré, heuristic fruitfulness always acts as a kind of realist indicator. This rider must be entered because he accepted non-constructive apriori principles which can be assessed, not in isolation from other assumptions, but within a wider nexus, where the latter might be a highly confirmed scientific system or some mathematical theory yielding answers to hitherto undecided questions.

The parallel between Kant and Poincaré is striking. We know that Kant ascribed a regulative role to the Ideas both of God and of Nature conceived as the infinite totality of physical events, which is why scientists can legitimately act on the assumption that the unity of Nature is guaranteed by a Supreme Being. A true and unified theory reveals the presence of a Deity whose existence should however not be a postulate of the system, but a tentative conclusion drawn from, or rather about it. The 'as if' mode of thought is therefore an important component of Kantian philosophy. Poincaré similarly rejected the existence of actual infinities which, according both to him and to Kant, are bound to engender antinomies. He nonetheless accepted some propositions about transfinite numbers as a *façon de parler* about sets which, despite remaining finite, grow beyond every conceivable limit; and one has the impression that he was not prepared, in the name of philosophical purism, to sacrifice any part of 'useful' mathematics. This kind of pragmatism is illustrated by his attitude towards classical analysis, and more particularly towards the Axiom of Choice. Far from rejecting this postulate, he accepted it, together with the Induction Scheme and the Axiom of Reducibility, as a kind of synthetic a priori truth.

Poincaré objected to the impredicative character of Zermelo's *proof* rather than to the use made in it of the Axiom of Choice. He had similarly criticized Russell's definition of the set ω of natural numbers as the intersection of all the inductive classes; for ω itself is an inductive set, so that the Vicious Circle Principle (VCP) is violated. When Russell finally developed his theory of ramified types, Poincaré called upon him to acknowledge the synthetic apriori character of the axioms of Infinity and of Reducibility.

Thus he adhered neither to formalism nor to strict constructivism but to a form of Kantianism. On the one hand, he held that abstract objects should, whenever possible, be *constructed* in subjective time. The latter must therefore be invoked in order to legitimize the iteration of any process needed to generate a basic mathematical entity. (It should be noted that his requirement does not entail the absoluteness of *physical* time.) On the other hand, Poincaré stressed the role of intellectual intuition in underwriting some a priori principles which go far beyond deductive logic. In this way, he arrived at the primacy of arithmetic, and more particularly of the Principle of Induction, over all logical principles; for all formal systems—including the purely logical ones—must be grasped recursively and hence arithmetically. This viewpoint was to be defended—much later—by Paul Bernays (Bernays 1976b).

It must nonetheless be admitted that Poincaré's overall position, which tries to reconcile synthetic opriorism with constructivism, cannot be pronounced a success; for he appears not to have realized that a choice had to be made between the sacrifices demanded by constructivism on the one hand, and on the other, the integrity of a classical mathematics bolstered up by a priori principles.

NOTE: *In the present chapter, the more familiar symbols for material implication and bi-implication, namely $\rightarrow$ and $\leftrightarrow$, will be used instead of $\Rightarrow$ and $\Leftrightarrow$. The latter are used in the remainder of the book in order to avoid confusion with the sign $\rightarrow$, which describes various limiting processes both in analysis and, more generally, in topology. No such risk of confusion arises in the present chapter.*

5.2 Mathematical Induction

Let us now carefully examine some of Poincaré's theses about the foundations of mathematics. In his 1902 he takes the view that mathematics is creative in at least two senses of the word: on the one hand, all the fundamental axioms cannot be logically true; on the other, the rules of inference used in mathematics are not exclusevily 'those of the syllogism'. In a more modern terminology, it can be said that according to Poincaré, mathematical reasoning could not possibly be entirely deductive; otherwise, all theorems would be implicitly contained in the axioms and would thus provide no genuinely new information; a consequence which he found unacceptable, not to say absurd.

> What is the nature of mathematical reasoning? Is it really deductive, as is commonly supposed? Careful analysis shows us that it is nothing of the kind, that it participates to some extent in the nature of inductive reasoning, and for that reason it is fruitful. But none the less does it retain its character of absolute rigour; and this is what must first be shown. (Poincaré 1902, p. xxiv)

> Finally, if the science of number were merely analytical, or could be analytically derived from a few synthetic intuitions, it seems that a sufficiently powerful mind could with a single glance perceive all its truths; nay, one might even hope that some day a language would be invented simple enough for these truths to be made evident to any person of ordinary intelligence.
>
> Even if these consequences are challenged, it must be granted that mathematical reasoning has of itself a kind of creative virtue and is therefore to be distinguished from the syllogism. (Poincaré 1902, p. 3)

Concerning the allegedly deductive nature of all mathematical proofs, Poincaré wrote:

> At first, do you think these logicians have always proceeded from the general to the particular, as the rules of formal logic would seem to require of them? Not thus could they have extended the boundaries of science; scientific conquest is to be made only by generalisation. (Poincaré 1906a, p. 23)

Following Kant, Poincaré held that among the mathematical principles there must be at least one which is not analytic but synthetic apriori. Mathematical induction, which is captured by the scheme: $P(0), (P(n) \rightarrow P(n + 1)) \,/\, (\forall_n)P(n)$ was chosen as the candidate *par excellence* for this elevated status. We ought therefore to analyze this rule of inference in some detail.

Poincaré's assessment of the status of induction is somewhat ambiguous. Since nobody seriously doubted its necessary character, it had somehow to differ from physical induction. Unlike its empirical counterpart, the 'principle of mathematical recurrence' was taken to relate not to any external reality, but to the process through which the mind infallibly knows itself capable of generating the natural number sequence. This account parallels Kant's well-known view that arithmetic expresses our intuition of succession in pure time. In the Prolegomena, Kant wrote:

> Mathematical Judgments are all synthetic. This fact seems hitherto to have altogether escaped the observation of those who have analyzed human reason . . .
>
> Now, the intuitions which pure mathematics lays at the foundations of all its cognitions and judgments, which appear at once apodeictic and necessary, are space and time. For mathematics must first present all its concepts in intuition, and pure

mathematics in pure intuition, i.e. it must construct them. . . . Geometry is based upon the pure intuition of space. Arithmetic attains its concepts of numbers by the successive addition of units in time, and pure mechanics especially can attain its concepts of motion only by employing the representation of time. (Kant 1953, §§2,8)

This quotation should be compared with the following one taken from Poincaré's 1902:

> We cannot therefore escape the conclusion that the rule of reasoning by recurrence is irreducible to the principle of contradiction. Nor can the rule come to us from experiment . . .
>
> This rule, inaccessible to analytical proof and to experiment, is the exact type of the *à priori* synthetic intuition. On the other hand, we cannot see in it a convention as in the case of the postulates of geometry. (Poincaré 1902, pp. 12–13)

This passage demonstrates that Poincaré held a Kantian view of arithmetic while simultaneously rejecting Kant's synthetic apriori conception of geometry; for in the case of geometry, we have a great latitude in the choice of the appropriate axioms, which must therefore be freely adopted conventions.

In order to account for the synthetic character of mathematical induction, Poincaré begins by looking upon it as an infinite, or rather as an indefinite, sequence of syllogisms condensed into one scheme. In other words, the above rule of inference ought to be treated as shorthand for: $P(0), [P(0) \rightarrow (P(1)], ..., [P(n-1) \rightarrow (P(n)] / P(n)$, for every natural number n; when interpreted in this way, mathematical induction can however achieve some 'economy of thought' but certainly no real increase in the content of its premises; for by Modus Ponens and hence by logic alone, we can, for every given n, pass from

$[P(0) \wedge ([P(0) \rightarrow (P(1)] \wedge ... [P(n-1) \rightarrow (P(n)]]$ to $P(n)$.

> The essential characteristic of reasoning by recurrence is that it contains, condensed so to speak in a single formula, an infinite number of syllogisms . . .
>
> But however far we went, we should never reach the general theorem applicable to all numbers, which alone is the object of science. To reach it we should require an infinite number of syllogisms and we should have to cross an abyss which the patience of the analyst, restricted to the resources of formal logic, will never succeed in crossing. (Poincaré 1902, pp. 9–11).

In the end, Poincaré saw the solution to the problem of mathematical creativity in that some principles like the inductive scheme proceed from the particular to the general, thereby increasing the content of their premises. On this view, there should be a deep analogy between physical induction on the one hand, and proofs by 'recurrence' on the other; the only difference being that unlike the former, the latter are infallible.

> It cannot escape our notice that here is a striking analogy with the usual processes of induction. But an essential difference exists. Induction applied to the physical sciences is always uncertain, because it is based on the belief in a general order of the universe, an order which is external to us. Mathematical induction—proof by recurrence—is, on the contrary, necessarily imposed by us, because it is only the affirmation of a property of the mind itself . . .
>
> No doubt, mathematical recurrent reasoning and physical inductive reasoning are based on different foundations, but they move in parallel lines and in the same direction—namely from the particular to the general. (Poincaré 1902, pp. 13–14)

Note that by accepting—*en bloc*—mathematical induction as a synthetic a priori principle, Poincaré obviates the difficulty of regarding it as an *actual infinity* of syllogisms. Thus synthetic apriorism will, time and again, enable Poincaré to mitigate the unreasonable demands of constructivism.

> To prove even the smallest theorem he must use reasoning by recurrence, for that is the only instrument which enables us to pass from the finite to the infinite . . . In this domain of arithmetic we may think ourselves very far from infinitesimal analysis, but the idea of mathematical infinity is already playing a preponderant part, and without it there would be no science at all, because there would be nothing general. (Poincaré 1902, p.11)

In his review of Poincaré's 1902, Russell pointed out that mathematical induction in no way enables us to generalize; for the premise $[P(n) \rightarrow P(n+1)]$, in which the free variable n occurs, actually stands for $(\forall n)[P(n) \rightarrow P(n+1)]$. In a proof by recurrence, we therefore pass from one *universal* proposition to another, namely from

$(\forall n)[P(0) \wedge ((P(n) \rightarrow P(n+1))]$ to $(\forall n)P(n)$,

so that any similarity between physical and mathematical induction vanishes. Russell moreover maintained that the set ω of natural numbers is *defined* by means of the inductive scheme; that is: ω is to be regarded as the smallest inductive class or as the intersection of all classes X such that $[(0 \in X) \wedge (\forall y \in X)((y+1) \in X)]$, where $(y+1) =_{\text{Def.}} y \cup \{y\}$. *By definition*, ω would thus satisfy the principle of induction.

Russell's first objection seems to me ungainsayable; but the second turned out to be far from cogent, especially after the breakdown of the logicist programme. In his reply, Poincaré appears not to have understood, or rather not to have wanted to understand Russell's first argument. He interpreted it as meaning that the induction principle is more general than its conclusion: that the *rule*, $P(0), (\forall n)[P(n) \rightarrow (P(n + 1)] / (\forall n)P(n)$, is somehow more general than the *formula* $(\forall n)P(n)$. This reply both distorts Russell's position and constitutes either a truism or else a category mistake. It is astonishing that Poincaré did not choose, instead, to attack Russell's second contention. He could have remarked that by *defining* ω in purely logical terms, Russell had merely shifted the difficulty; for he had now to demonstrate that in his *proofs* of the standard classical theorems, only logical postulates are invoked. We know that Russell was subsequently compelled to resort to axioms such as those of Infinity and of Reducibility which, by his own admission, are not logical in character. Thus some mathematical axioms are synthetic after all. Should they moreover be considered true independently of experience, then they must be accepted as synthetic apriori propositions. Despite the cogency of such a reply, there is the strong presumption that in 1902, Poincaré would have rejected this whole line of reasoning, or rather that he would have found it wanting; for it implicitly concedes that mathematical proofs are purely deductive and hence that the whole content of a mathematical theory is captured by its axioms, a consequence which Poincaré refused to accept. Despite Russell's justified criticisms, he therefore went on claiming that mathematical induction proceeds from the particular to the general, thus increasing the content of its premises:

> . . . scientific conquest is to be made only by generalisation. In one of the chapters of 'Science and Hypothesis' I have had occasion to study the nature of mathematical reasoning, and I have shown how this reasoning, without ceasing to be absolutely rigorous, could lift us from the particular to the general by a procedure I have called mathematical induction. (Poincaré 1906a, p. 23)

Poincaré thus viewed induction as the essential mechanism of progress in mathematics as well as in the natural sciences. In the same way that various conventions provide physics with factually empty classificatory schemes, neither the logical axioms nor the deductive rules of inference convey any genuine mathematical information. In both cases, only 'inductive generalisations' widen the scope of our knowledge.

In his philosophy of mathematics and more particularly in his appraisal of induction, Poincaré thus appears to have committed a series of logical blunders. But in this case as in many others, it proves far more rewarding to reconstruct his position rationally, while still appealing to views expressed in some of his writings.

> Why is this view [about mathematical induction] imposed upon us with such an irresistible weight of evidence? It is because it is only the affirmation of the power of the mind which knows it can conceive of the indefinite repetition of the same act, when the act is once possible. The mind has a direct intuition of this power, and experiment can only be for it an opportunity for using it and thereby of becoming conscious of it. (Poincaré 1902, p. 13)

Poincaré's position can be reconstructed as follows. Like Kant, he regards the natural numbers as being continually created by acts carried out in the pure intuition of time. Since time is the subjective form of our inner sense, it would be meaningless either to affirm or to deny that such a pure form has an absolute beginning. The passage from any integer n to its successor n + 1 is therefore indiscernible from the 'first' step taking us from 0 to 1. The transition from one number to the next can thus be recognized as an indefinitely repeatable operation; so we have to do with a singular act which nonetheless epitomizes a generally applicable rule. In other words: through intuiting one process of succession in time, the mind apprehends it as typical of the successive steps giving rise to the natural number sequence. We can thus effect a transition from the particular to the general which, being irreducible to any analytical principle, accounts for the profound analogy between the mathematical and the physical inductive schemes. There remains however one important difference: the pure character of subjective time guarantees the necessity of all propositions obtained by mathematical recurrence, but as soon as anything empirical intervenes into the temporal process, we have to do with physical inductive inferences yielding only contingent laws. The parallel between Kant's and Poincaré's viewpoints is therefore astonishingly close. The Kantian recourse to temporal considerations became even more explicit when Poincaré later tried to find a remedy to the logical paradoxes (see below, 5.5).

5.3 The Logical Paradoxes

Poincaré initially spoke of the sterility and even of the dead hand of symbolic logic which threatens to reduce mathematics to one vast tautology. He then sarcastically admitted to having been mistaken: far from being sterile, the new 'Logistic' engendered a plethora of *antinomies*. In discussing these, it will suffice to examine two typical paradoxes, namely those of Russell and of Berry, which occupy the opposite ends of a whole spectrum.

Berry considers the set E of all natural numbers nameable in less than one hundred words. E is obviously finite. Hence there exist numbers larger than every member of E; and among them, there must be a smallest element, α say. Since α exceeds every member of E, $\alpha \notin E$. But α is also denoted by the expression 'the smallest natural number greater than every natural number nameable in less than one hundred words' which contains less than one hundred words. Hence, by the definition of E, $\alpha \in E$. There follows a contradiction, called a semantic or linguistic paradox. (Note: I have slightly altered Berry's paradox—as normally presented—by considering not the first non-element of E but the smallest number greater than every member of E. This does not however alter any argument for, or against the existence of α).

As for Russell's paradox, it arises from the definition of the set $\underline{a}$ as the collection of all objects which do not belong to themselves. (For any non-set z, we stipulate that $(\forall y)(y \notin z)$; which entails $z \notin z$.) Thus:

(1) $\underline{a} =_{\text{Def.}} \{x: x \notin x\}$.

Therefore:

(2) $(\forall x)[(x \in \underline{a}) \leftrightarrow (x \notin x)]$.

Through substituting $\underline{a}$ for x, we obtain the logical falsehood:

(3) $[(\underline{a} \in \underline{a}) \leftrightarrow (\underline{a} \notin \underline{a})]$.

Note that (3) is self-contradictory because it is false for both truth-values of $\underline{a} \in \underline{a}$. Russell's paradox, as opposed to Berry's, is often said to be of the logico-mathematical type. Russell however denied the legitimacy of any fundamental distinction being made between *different kinds* of paradox. He furthermore proposed what is, to my mind, the best approach to the elimination of *all* the antinomies. It consists in surveying the extant paradoxes and in then identifying, in all of them, the trait prima facie responsible for their self-contradictory character. Since I have neither enough space nor enough knowledge to reconstruct all of Russell's analyses, I shall list some of his methodological principles, most of which were incidentally shared by Poincaré. I shall subsequently examine how these principles bear on Russell's solution to some of the notorious antinomies like the ones mentioned above.

[A] Epistemological Fundamentalism (or Integralism)

This is how G. Heinzmann characterizes the directive enjoining us to remove the paradoxes not only from logic and mathematics, but also from all areas of human knowledge. The French noun *intégrisme* conveys the meanings of both fundamentalism and holism (see Heinzmann 1985, Chapter 4). In accordance with Russell's heuristic, we should try to identify one aspect which is common to all the linguitic and logical paradoxes; which gives rise, in every case, to some antinomy; and which could be neutralized by the application of one intuitively self-evident rule. Let us henceforth denote this latter—as yet unknown—therapeutic principle by the letter X.

Russell's fundamentalism flows directly from his logicism. Since mathematics belongs to logic, all mathematical contradictions must be traced back to some defect either in the logical axioms or in the deductive rules of inference. Mathematics cannot be isolated in such a way that through tampering with the specifically mathematical postulates, we rid ourselves of the specifically mathematical paradoxes. According to Russell, semantics is moreover part and parcel of logic, so that the semantic antinomies must have a strictly logical character.

Poincaré's fundamentalism is more difficult to account for. He believed in the primacy of mathematics and more particularly of arithmetic, over all other forms of knowledge. He implicitly held that in order to be comprehensible, a formal system must be based on some recursive grammar—on a grammar built on successive constructions in pure time. Thus all systematic thought possesses an *arithmetical* structure, so that any error in the foundations of mathematics would have repercussions throughout all other branches of knowledge. Conversely, a remedy to the mathematical paradoxes might well abolish them everywhere. Even if these arguments be found somewhat unconvincing, fundamentalism remains closely linked to Poincaré's rule which requires every solution to any problem to be nonadhoc. Other things being equal, the more universal the solution, the less ad-hoc it is likely to be (see [c] below).

[B] The Integrity of Mathematics

X must be such that even after its adoption, we will still be in a position to derive the whole of known mathematics from self-evident axioms. X should therefore not be too restrictive. Russell remained bound by this rule and Poincaré implicitly subscribed to it, especially with regard to the 'useful' parts of mathematics.

Whitehead and Russell eventually succeeded in setting up a system satisfying condition [b]. Their *Principia*, which are based on the ramified theory of types, reconstruct classical mathematics while blocking the known routes leading to the paradoxes. By contrast, Poincaré's *philosophical* insights, profound though they are, do not add up to an integrated system which retrieves the whole of extant mathematics. Poincaré furthermore failed to acknowledge the obvious sacrifices demanded by his *quasi-intuitionstic* approach; all of which does not detract from his having exposed serious flaws in Russell's *philosophical* logic. This is why Russell and Whitehead were the real mathematical doers while Poincaré played the role of an intuitive philosopher. (For a more comprehensive treatment, see Kilmister 1984, Chapters 3 and 4).

[c] Banning Adhoc Stratagems

Poincaré and Russell held that no solution to any foundational problem should be adhoc, that is, tailored to the immediate needs of some cause. In the present case, the cause is that of reconstructing mathematics while evading the antinomies. According to Poincaré's general methodology, every theory should be a unified thesis which not only subsumes a number of known facts, but also yields new predictions or else establishes unexpected relations between hitherto disconnected domains. The scientist ought therefore to refrain from arbitrarily augmenting the number of independent hypotheses, more particularly, from appealing to as many laws as there are facts to account for. As seen in Chapter 4 above, Poincaré had criticized some adhoc conjectures like the Fitzgerald contraction hypothesis, whose only role was to explain *away* the repeated refutations of the ether programme. He therefore agreed with Russell about searching for a unique principle X which would eliminate *all* the paradoxes, whether these be logical or linguistic. He also insisted on the major role which *intuition* must play in all mathematical disciplines; but he gave this word different meanings among which at least two must be distinguished. First of all, he underlined the *heuristic* importance of intuition while mocking at Couturat's contention that symbolic logic would "give new wings" to discovery. Let us call this sort of intuition *creative* without giving any very precise meaning to this term. When Russell stumbled on the paradoxes and then embarked on casting the foundations of mathematics anew, Poincaré predicted that an appeal to some form of intuition would have to be made. Russell agreed, while insisting that this second form of intuition should be of the strictly *logical* kind; and he took Poincaré's side against Couturat by conceding that the task of logic was not to aid mathematical invention but to analyse thought into its primitive elements. Among the latter, the following must figure: a basic vocabulary including the logical connectives, a set of formation and of inference rules, and an adequate system of axioms. Being presupposed by all explanations, these elements are intrinsically inexplicable. We nonetheless have a *logical intuition* which may grant or else refuse its assent to any proposed first principles. In this spirit, Russell was later rightly to maintain that type theory was consonant with logical commonsense. On this point of principle, Frege, Russell, and Poincaré were all at one. Were it not for this large overlap between Poincaré's and Russell's methodologies, then any dialogue between them would anyway have been fruitless. At this point however, we also encounter some major difficulties, for when it comes to assessing the status of such axioms as those of Infinity, of Choice, or of Reducibility, our logical intuition may either be neutral or else remain silent. It is furthermore possible for the *creative* and the *logical* intuitions to pull in opposite directions. A formal argument is meant to display *all* the primitive elements underwritten by 'logical' intuition; it consequently puts on a par the 'creative' moves and the 'routine' procedures carried out in a deduction. But a proof generally possesses a nerve centre which would be concealed by any such exclusively formal rendition. An essential function of *creative intuition* is either to omit or else to compress into a single move all the trivial steps: that is: all the obvious but long and tedious combinations of primitive elements. An important aspect of a mathematician's talent consists in his ability to neglect the routine components of a planned proof and—*before* embarking on the deduction—to anticipate the difficult steps which are the only obstacles to be overcome. Once this is done, the proof might well become a matter of mere verification, of filling the 'obvious' gaps, which in no way prevents mistakes about the obviousness of these steps from being made. Poincaré was thus acutely aware that a formal argument would eliminate the important disparities between different moments of one and the same proof. As for *logical* intuition, Russell looked upon it as being both indispensable and unreliable; for in a case like that of the Comprehension Scheme, it leads us straight into contradictions (see Scheme (κ) below). But where a system is presumed consistent, intuition remains the only recourse against adhocness; it is more particularly needed as independent support for the principle X which, as already mentioned, must satisfy the further requirement of avoiding the paradoxes while enabling us to reconstruct classical mathematics. Gödel was later to remark that within mathematics, this kind of intuition plays a role analogous to that of empirical evidence in the natural sciences. For example, in modifying the

Comprehension Scheme, we judge what extensions can safely be held to exist in a way similar to that in which the existence of material objects is intuited. There can however be no question of our logical intuition being in any way infallible. Let us note that in the physical sciences, as soon as we transcend the phenomenological sphere and start referring to some external reality, we also renounce infallibility and have to content ourselves with conjectural knowledge. According to Gödel, the same applies to hypotheses about entities like sets and classes (Gödel 1983a).

With regard to logical intuition, let us conclude that Poincaré and Russell largely agreed about the necessity of finding an integral and non-adhoc solution to *all* the antinomies. We shall now see how these methodological principles informed Poincaré's appraisal of Zermelo's system.

Poincaré held that like any mathematical system, set theory should fulfil two conditions: its basic postulates must be consistent and they moreover ought to be intuitively self-evident. With regard to the consistency condition: given the supposedly fundamental character of Zermelo's system, it is impossible to determine a model of all its axioms without invoking at least one of them, so it looks as if one could at best hope to find a *relative* consistency proof for set theory. There remains however one other *prima facie* possible solution. Poincaré *intuitively* identified a fundamental characteristic of all axiomatisable theories S, namely that of being recursively enumerable. Hence it appeared as though it were in principle feasible to design a machine which would churn out all the theorems of S in the form of one potentially infinite proof. Let D denote such a proof. According to Poincaré, we could establish the consistency of S by showing that if no contradiction occurs in D at some stage n, then the same will hold for stage n + 1, where n is an arbitrary natural number. In other words, let D consist of the sequence of sentences, A_0, A_1,... We could try to prove that if no formula of the form $B \wedge \neg B$ occurs among $A_0, \ldots, A_n$, then the same will hold good for $A_0, \ldots, A_n, A_{n+1}$. Given Gödel's theorems, we now know that if S consistently yields ordinary arithmetic, then no such proof is possible. But independently of this result of which he was naturally unaware, Poincaré pointed out that Zermelo could not legitimately avail himself of such an *inductive* method. Since one of the aims of set theory is to provide a *foundation* for mathematical induction, the latter should not be presupposed in any argument for the cogency of set theory. (Poincaré was later to make the same objection to Whitehead's and Russell's *Principia*.)

There remains the question whether the set-theoretic postulates might not, all told, be self-evident. Poincaré was however quick to expose the adhoc character of Zermelo's axioms. He showed that they were artificially *doctored* to avoid the paradoxes while leaving certain mathematical proofs intact. Because it was tailored to the narrow purpose of preserving Cantor's paradise, set theory in effect lacked a unifying philosophical principle.

In what follows, we shall often need to refer to the Zermelo-Fraenkel system which will henceforth be denoted by (ZF). It is therefore worth our while to formulate (ZF) in its most general form, that is: in a form which allows for the existence of *Urelementen*, or of entities which are not classes.

(ZF) has two primitive descriptive terms: the monadic predicate s [$s(x) \equiv$ x is a set] and the binary relation $\in$ [$(x \in y) \equiv$ x belongs to the set y].

The following abbreviations will prove useful:

$(x \subseteq y) \equiv_{\text{Def.}} (\forall z)[(z \in x) \rightarrow (z \in y)]$;

$(x \subset y) \equiv_{\text{Def.}} [(x \subseteq y) \wedge (x \neq y)]$;

$(x = \{y,z\}) \equiv_{\text{Def.}} [s(x) \wedge (\forall v)((v \in x) \leftrightarrow ((v = y) \vee (v = z)))]$;

$(a \in \{y,z\}) \equiv_{\text{Def.}} (\exists x)[(x = \{y,z\}) \wedge (a \in x)]$; $(\{y,z\} \in a) \equiv_{\text{Def.}}$ $(\exists x)[(x = \{y,z\}) \wedge (x \in a)]$;

$(x = <y,z>) \equiv_{\text{Def.}} [s(x) \wedge (\forall v)((v \in x) \leftrightarrow ((v = \{y,y\}) \vee (v = \{y,z\}))]$;

$(a \in <y,z>) \equiv_{\text{Def.}} (\exists x)[(x = <y,z>) \wedge (a \in x)]$; $(<y,z> \in a) \equiv$ $(\exists x)[(x = <y,z>) \wedge (x \in a)]$;

$\mathbf{R}(f) \equiv_{\text{Def.}} [s(f) \wedge (\forall x \in f)(\exists y,z)(x = <y,z>)]$; (Intuitively: f is a relation).

$\mathbf{F}(f) \equiv [\mathbf{R}(f) v (\forall y,z,v)(((<y,z> \in f) \wedge (<y,v> \in f)) \rightarrow (z = v))]$; (f is a function).

$\mathbf{F}_1(f) \equiv [\mathbf{F}(f) \wedge (\forall y,z,v)(((<z,y> \in f) \wedge (<v,y> \in f) \rightarrow (z = v))]$; (f is a one-one function).

$(x \approx y) \equiv [s(x) \wedge s(y) \wedge (\exists f)(\mathbf{F}_1(f) \wedge$

$(\forall z)((z \in x) \leftrightarrow (\exists v)(<z,v> \in f)) \wedge (\forall z)((z \in y) \leftrightarrow (\exists v)(<v,z> \in f))]$; (The set x is equinumerous with the set y iff there exists a one-one map f of x onto y.)

$\text{Ind.}(x) \equiv [s(x) \wedge (\forall y \in x)s(y) \wedge (\exists z \in x)(\forall v)(v \notin z) \wedge$

$(\forall z \in x)(\exists v \in x)(\forall a)((a \in v) \leftrightarrow ((a = z) \vee (a \in z)))]$; (x is an inductive set iff x is a set of sets such that the empty set belongs to x and, for any element z of x, the successor $z \cup \{z\}$ of z also belongs to x.)

$\varpi_0(x) \equiv [s(x) \wedge (\forall y)(\text{Ind.}(y) \rightarrow (x \in y))]$; ($\varpi_0(x) \equiv$ x is a natural number, i.e. x is a set which belongs to the intersection of all inductive sets y.)

$\varpi(x) \equiv [s(x) \wedge (\exists y)(\varpi_0(y) \wedge (y \approx x))]$; ($\varpi(x) \equiv$ x is a finite set, i.e. x is equinumerous with a finite number y.)

Turning to the set-theoretic axioms, let us start by pointing out that the following are not all mutually independent.

The first two postulates of (ZF) tell us that only sets can have elements and that sets are purely extensional entities—entities uniquely determined by their members:

[1] $(\forall x)(\forall y)[(x \in y) \rightarrow s(y)]$.

[2] $(\forall a,b)[(s(a) \wedge s(b) \wedge (\forall x)((x \in a) \leftrightarrow (x \in b))) \rightarrow (a = b)]$ (Axiom of Extensionality).

Apart from the extensionality postulate which flows directly from the intuitive concept of a *class*, only the following Comprehension Scheme:

(κ) $(\exists b)[s(b) \wedge (\forall x)((x \in b) \leftrightarrow \chi(x))]$,

where $\chi(x)$ is an arbitrary formula, appears to be intuitively self-evident; but it is well-known that Axiom (κ) is the one mainly responsible for the existence of the antinomies. For example, taking $\chi(x)$ to be $(x \notin x)$, we immediately infer Russell's paradox; which is why Zermelo proceeded to postulate those special cases of (2) which, apart from being presupposed in standard proofs, appeared to give rise to no obvious contradictions. For example he assumed:

[3] The existence of the empty set: $(\exists b)[s(b) \wedge (\forall x)((x \in b) \leftrightarrow (x \neq x))]$; i.e. $(\exists b)[s(b) \wedge (\forall x)(x \notin b)]$.

Moreover, for any two objects u and v, he postulated the existence of a set having u and v as its only members:

[4] $(\forall u,v)(\exists b)[s(b) \wedge (\forall x)((x \in b) \leftrightarrow ((x = u) \vee (x = v)))]$ (Pairing Axiom).

Axioms **[3]** and **[4]** are special cases, albeit very weak ones, of Scheme (κ). Much stronger instances of (κ) are needed in order to reconstruct ordinary mathematics. Cantor had already noticed that the classes involved in the set-theoretic paradoxes were very large; he had consequently described them as *coastless* or *frontierless multiplicities (uferlose Mannigfaltigkeiten).* Such are the classes of all ordinals, of all cardinals and of all sets not belonging to themselves. This is why Zermelo embarked on a huge operation of what Lakatos called 'monster-barring' (see Lakatos 1976,

p. 86). By fiat, Zermelo excluded from the family of respectable sets some collections which were considered 'excessively large'; yet neither he nor Cantor could point to any logical connection between the antinomies on the one hand, and the cardinality of the banned classes, on the other. Large size and inconsistency were simply concomitant features of some, but by no means all the paradoxes.

The essential defect of monster-barring is that it eliminates the *symptoms* but not the deep causes of a given illness. Russell had actually drawn Poincaré's attention to the fact that the sets implicated in Berry's paradox, far from being 'too large', were not even infinite; and Poincaré had acknowledged the correctness of Russell's criticism. Berry's paradox was however of a type which was of interest to very few mathematicians; the others dubbed it a mere linguistic antinomy, which was but another monster-barring move. With respect to some but not all of his postulates, Zermelo thus adopted a strategy called the principle of limitation of size. Russell himself had envisaged this approach and then rejected it on the ground that it generates only adhoc solutions. Limitation of size led Zermelo to his now well-known

[5] Separation Scheme (*Aussonderungsaxiom*):
$(\forall z)[s(z) \rightarrow (\exists b)(s(b) \wedge (\forall x)((x \in b) \leftrightarrow ((x \in z) \wedge E(x))))]$; where $E(x)$ is any wff (well-formed formula). (Intuitively, this axiom aserts that the elements of a set z which possess a well-defined property E also form a set.)

A. Fraenkel later realized that the Axiom of Separation was too weak to permit the development of a satisfactory set theory. Through his Replacement Scheme, he accordingly readmitted certain classes as legitimate sets. This scheme moreover turns out to be stronger than the *Aussonderungsaxiom*:

[6] Replacement Scheme (*Ersetzungsaxiom*):
$(\forall z)[(s(z) \wedge (\forall x \in z)(\exists^* y)C(x,y)) \rightarrow (\exists b)(s(b) \wedge (\forall y)((y \in b) \leftrightarrow (\exists x \in z)C(x,y)))]$; where, for any wff $A(y)$, $(\exists^* y)A(y)$ stands for $(\forall y,v)[(A(y) \wedge A(v)) \rightarrow (y = v)]$. (Intuitively: if z is a set such that, for any x in z, there exists at most one y satisfying $C(x,y)$, then the image b of z under C is a set.)

Note that **[6]** is also warranted by the principle of limitation of size: the image b of the set z under the partial function C can have no greater number of elements than z; since z is a set, that is to say a 'small' class, it follows that b is also small and hence a set.

Axioms **[5]** and **[6]** obviously constitute special cases of the Comprehension Scheme. They are introduced not on account of their intrinsic plausibility, but merely because they permit the derivation of results judged to be desirable. The system of postulates so far introduced however falls short of its main task, namely that of shoring up Cantor's paradise. With the exception of **[3]** and **[4]**, the axioms merely assert that if z is a set, then every collection having fewer members than z is also a set; but they do not tell us what z 's are admissible in the first place, nor do they enable us to increase the cardinality of any given z. This is why Zermelo resorted to a policy of reflation designed to augment the size of any given class. Thus he postulated the existence of the union of any set of sets:

[7] <u>The Existence of the Union:</u>
$(\forall z)[(s(z) \wedge (\forall y \in z)s(y)) \rightarrow (\exists(b)(s(b) \wedge (\forall v)((v \in b) \leftrightarrow (\exists a)(v \in a \in z)))]$; (the elements of elements of any set z constitute a set b.)

The set b just mentioned—which is usually denoted by $\cup z$—contains, and hence is more extended than every element of z. This axiom does not however go far enough; for in many mathematical proofs, we have to refer to collections of subsets of a given set. For example, in certain algebraic theories, we need to consider whole families of subgroups of a group or of ideals of a given ring. Zermelo consequently decreed that the totality of subsets of any set z constitutes another set which is usually denoted by *P*(z). Thus he implicitly assumed that since z is not 'coastless', *P*(z) will still be small enough to fall within the universe of sets. In other words, he postulated:

[8] <u>The Power Set Axiom</u>:
$(\forall z)[s(z) \rightarrow (\exists b)(s(b) \wedge (\forall x)((x \in b) \leftrightarrow (s(x) \wedge (x \subseteq z))))]$; (The subsets of a given set z constitute a set b, which will be denoted by *P*(z).)

To postulates **[1]**—**[8]**, the following <u>Axiom of Foundation</u> was later added:

[9] $(\forall x)[(s(x) \wedge (\exists y)(y \in x) \wedge (\forall y \in x)s(y)) \rightarrow (\exists z \in x)(\forall y \in x)(y \notin z)]$; (Every non-empty set of sets x possesses a member z whose intersection with x is empty).

By virtue of Cantor's theorem, we know that $|z| < |P(z)|$. But in going from z to *P*(z), the cardinality of z not only increases, but it does so in an uncontrollable way; for we have no clear idea of how large *P*(z) might be. With total impunity, we can deny or affirm that if z is infinite, $|P(z)|$ is the cardinal immediately above |z|. In other words, we have no intuitive grasp of what the universal quantifier in the expression 'all the subsets of z' really means.

Zermelo had in effect applied a method which Lakatos later called 'proof analysis', but Zermelo used this heuristic device in an entirely adhoc way (Lakatos 1976, pp. 42–56). Whereas Lakatos pointed out that a theorem applies *only* to the objects possessing the properties implicitly presupposed in a given proof, Zermelo boldly postulated the existence of all the entities needed in order for certain *global* results to remain valid. More precisely, Zermelo set out to posit entities which were meant to provide a framework for the whole of mathematics, namely those well-behaved classes which constitute respectable sets. Unlike other mathematicians, he could not rely on an *as yet unavailable* set theory in order to circumscribe the domain of discourse of his deductions. He was for example in no position to maintain that the well-ordering theorem holds within all *sets* in the same way that unique factorisation can be claimed to hold *in all Euclidean domains:* by using set theory, we can characterize *all* the Euclidean rings of which we anyway have many examples; but we do not know of any conditions both necessary and sufficient for a class to constitute a set in order then to claim that the well-ordering theorem holds within the class of all sets. To the extent that he refused to give up classical set theory, Zermelo found himself compelled to *postulate* what he needed in order for certain proofs to go through; which is precisely the reason why Poincaré objected to Cantorian set theory.

In no way do I wish to suggest that axiomatic set theory *does not work.* On the contrary, at the level of mathematical practice, it represents a great success while the type-theoretic systems which sprang from Frege's, Russell's, and Poincaré's insights proved so cumbersome as to be practically unusable. Especially in its impredicative Gödel-Bernays version, set theory remains one of the most powerful instruments at the mathematician's disposal. It gives him the whole of useful mathematics and much else besides. It is furthermore expressible in a first-order language whose logic is complete; but as already mentioned, we are not dealing here with actual mathematical practice but with a philosophical critique of foundations. Especially after Cohen's results, we have come to realize that Russell's and Poincaré's criticisms were justified. Though still believing in a Platonic universe of sets, Gödel himself conceded that we do not as yet possess a clear intuition of the notion of class. Thus the set-theoretic axioms may well turn out to be a laundry list whose foundations and validity are circumscribed by type-theoretic logic.

To sum up: except for **[1]** and **[2]**, all the axioms so far formulated are special cases of the Comprehension Scheme **[κ]**: each one of them asserts the existence of the extension b of some propositional function *f*(x). The Separation and Replacement Schemes obey the Principle of Limitation of Size whose connection with the antinomies remains unclear; the other axioms are introduced, in a purely adhoc manner, for the sole purpose of allowing certain proofs to go through. In some cases Zermelo *decrees* that objects sharing some property are entitled to form a 'syndicate', that is: to have a set as their extension. All other cases are passed over in silence. Poincaré consequently realized that *logical intuition* in no way supports these isolated instances of the Comprehension Axiom, but goes straight to the general scheme itself; which unfortunately lands the mathematician into the well-known antinomies. There ought therefore to be an appeal to some deeper intuition which corrects our prima facie unconditional acceptance of Scheme **[κ].**

Before going any further, let us formulate the <u>Axiom of Infinity</u> which is needed in order to develop classical analysis:

[10] <u>The Axiom of Infinity</u>: $(\exists x)$ Ind. (x). (There exists at least one inductive set x; thus x contains, by definition, the sequence of all natural numbers and is therefore infinite).

Poincaré treats the following postulate (AC)—called Axiom of Choice—with greater indulgence than he does Axiom [10]:

[11] <u>The Axiom of Choice</u> (AC):
$(\forall x)[(s(x) \wedge (\forall y \in x)(s(y) \wedge (\exists z)(z \in y)) \wedge (\forall y \in x)(\forall v \in x)((\exists t)((t \in y) \wedge (t \in v)) \rightarrow (y = v))) \rightarrow (\exists a)(s(a) \wedge (\forall t \in a)(\exists y \in x)(t \in y) \wedge (\forall y \in x)(\exists b)(\forall t)((t = b) \leftrightarrow ((t \in a) \wedge (t \in y))))]$. (For any set x of non-empty disjoint sets, there exists at least one set a such that: a is contained in the union of all members of x and, for every element y of x, the intersection $a \cap y$ of the two sets a and y is some singleton $\{b\}$.)

Poincaré seems prepared to accept the Axiom of Choice as a synthetic apriori principle; which reveals, once again, his deep intuitive grasp of the philosophical foundations of logic. We know that the Axiom of Choice is closely linked to classical logic in that it flows from our ordinary use of the indefinite article. Hilbert denoted this article by an ε-symbol which represents a universal choice-function governed by the following two postulates:

[12] $(\exists x)\mathbf{K}(x) \rightarrow \mathbf{K}(\varepsilon y \mathbf{K}(y))$, and

[13] $(\forall x)[\mathbf{K}(x) \leftrightarrow \mathbf{H}(x)] \Rightarrow [\varepsilon(y\mathbf{K}(y)) = (\varepsilon z\mathbf{H}(z))]$, where $\mathbf{K}(x)$ and $\mathbf{H}(x)$ are arbitrary wffs.

According to **[12]**, every property $(\lambda x\mathbf{K}(x))$ possessing a non-empty extension is true of the entity $(\varepsilon y(\mathbf{K}(y))$ which is selected by the universal choice functor ε from among all the elements of this extension. As for **[13]**, it tells us that ε is purely extensional; i.e. that $(\varepsilon y\mathbf{K}(y))$ depends only on the class $\{x\text{: } \mathbf{K}(x)\}$ and not on the (Fregean) sense of $(\lambda x\mathbf{K}(x))$. It is well-known that within the framework of this ε-logic, the 'Axiom' of Choice can be deduced from the remaining postulates of set theory, i.e. from **[2]**-**[8]**. It should however be understood that the underlying logic is thereby taken to be an ε-logic including Schemes **[12]** and **[13]**; moreover that the Replacement Scheme **[6]** now comprises *all* the wffs *C*(x,y*)*, whether or not they contain the ε-symbol. We therefore have an extended scheme which ought, strictly speaking, to be denoted by $\mathbf{[6]}_{\varepsilon}$.

The above considerations show that (AC) represents a quasi-logical principle; or rather that its excess content $\neg K \vee (AC)$ over the other postulates of set theory is a logical truth; where we put: $K \equiv$ conjunction of the postulates **[1]**–**[8]**. (Needless to say, K contains—as conjuncts—only those instances of $\mathbf{[6]}_{\varepsilon}$ which are actually invoked in the deduction of (AC) from the remaining axioms of set theory. By the very nature of a formal proof, these instances are finite in number).

For the sake of completeness, let us formulate the Generalised Continuum Hypothesis (GCH), which was to play a major role respectively in Gödel's and in Cohen's proofs of the relative consistency and of the independence of (AC)&(GCH):

[14] The Generalized Continuum Hypothesis (GCH): $(\forall x,y,z)[(s(x) \wedge s(y) \wedge s(z) \wedge \neg\overline{\varpi}(x) \wedge (\forall v)((v \in z) \leftrightarrow (s(v) \wedge (v \subseteq x))) \wedge (\exists u)(s(u) \wedge (x \approx u \subseteq y)) \wedge (\exists t)(s(t) \wedge (y \approx t \subseteq z))) \Rightarrow ((x \approx y) \vee (y \approx z))]$. (Given any three sets x, y, and z such that x is infinite, z is the power set of x, x is equinumerous with a subset of y, and y is equinumerous with a subset of z, then y must be equinumerous either with x or with z.)

As already mentioned, (ZF) denotes the formal system consisting of axioms **[1]**–**[10]**. (ZFC) and (ZFCH) will henceforth refer to the systems (ZF)&**[11]** and (ZF)&**[11]**&**[14]** respectively.

Poincaré's appraisal of the Axiom of Choice shows that he was far from being a strict constructivist. To repeat, he was essentially a Kantian whose constructivism did not go beyond that of Kant. He accepted not only the principle of the excluded middle but also a number of synthetic apriori principles which violate most modern forms of constructivism. We have also mentioned that independently of Zermelo's use of (AC) in his proof of the well-ordering theorem, Poincaré took exception to the very structure of this proof; for he rightly held it to be impredicative. It is thus high time for us to define the twin notions of predicativity and of vicious circularity. Poincaré maintained that there was vicious circularity whenever an entity a is defined in terms of a collection E some of whose elements depend, in turn, on a, more particularly: whenever E is taken to contain a among its members. He called any deduction involving a vicious circle impredicative. As for the Vicious Circle Principle (VCP), it asserts the invalidity of all impredicative arguments.

Zermelo's celebrated proof obviously violates the (VCP). For let m be any non-empty set and let f be a choice function on m. That is: $[(D(\mathrm{f}) = P(\mathrm{m})) \wedge (\forall \mathrm{x} \in P(\mathrm{m}))((\mathrm{x} \neq \emptyset) \rightarrow (\mathrm{f'x} \in \mathrm{x}))]$, where we define: $D(\mathrm{f}) =_{\mathrm{Def.}} \{\mathrm{x}: (\exists \mathrm{y})(\langle \mathrm{x,y} \rangle \in \mathrm{f})\}$, $R(\mathrm{f}) =_{\mathrm{Def.}} \{\mathrm{y}: (\exists \mathrm{x})(\langle \mathrm{x,y} \rangle \in \mathrm{f})\}$, and $\mathrm{f'x} =_{\mathrm{Def.}} \iota\mathrm{y}[\langle \mathrm{x,y} \rangle \in \mathrm{f}]$. Zermelo considers the family E of all subsets z of m such that there exists a well-ordering relation r on z with the property that $(\forall \mathrm{x} \in \mathrm{z})[\mathrm{f'(m} \backslash \mathrm{z_x}) = \mathrm{x}]$; where $\mathrm{z_x} =_{\mathrm{Def.}} \{\mathrm{v}: (\mathrm{v} \in \mathrm{z}) \wedge (\langle \mathrm{v,x} \rangle \in \mathrm{r})\}$. Zermelo then defines d as the union of all members of E, that is: $\mathrm{d} =_{\mathrm{Def.}} \cup \mathrm{E}$. He finally shows that $\mathrm{d} \in \mathrm{E}$ and that d = z. Thus d both belongs to E and is defined in terms of E (namely through $\mathrm{d} = \cup \mathrm{E}$). Zermelo's argument thus violates the (VCP) and must, according to Poincaré, be pronounced invalid.

Every paradox so far encountered not only involves, but also arises from some vicious circularity. In each case, some entity, despite being defined in terms of a totality which it ought to transcend, turns out to belong to that same totality. The first such antinomy was discovered by Burali-Forti, who considered the *ordinal* γ of the class Ω of *all ordinals*. Thus $\gamma \in \Omega$, so that despite being defined in terms of Ω, γ must also lie in Ω. It is then established that $\gamma \notin \Omega$ because γ exceeds *all* ordinals and hence every member of Ω; that is: γ must transcend Ω; whence a contradiction. In the case of Cantor's paradox, we similarly envisage the set K of *all cardinals* in terms of which a specific cardinal p is then defined. Thus $\mathrm{p} \in \mathrm{K}$. By virtue of Cantor's theorem, p can however be chosen so as to be strictly larger than every element of K; hence $\mathrm{p} \notin \mathrm{K}$, which yields yet another contradiction. As for Berry's antinomy, let E be the list of all natural numbers nameable in less than one hundred words. Since all dictionaries are finite, so is E. Hence there exist strict upper bounds of E from amongst which we select the smallest one, b say. Thus both $b \notin E$ and b is defined in terms of E. But by definition, b is *the smallest number greater than every number nameable in less than one hundred words* and is thus manifestly so nameable. By definition of *E,* we must therefore have $b \in E$. Once again, a contradiction arises from some object *b* being defined in terms of a totality *E* to which *b* itself both must and cannot belong.

Finally, Russell's set $\underline{a}$, which consists of all sets not belonging to themselves, *does not belong to itself;* i.e.

$$\underline{a} = \{x\colon s(x) \wedge (x \notin x)\} \notin \{x\colon s(x) \wedge (x \notin x)\} = \underline{a};$$

hence $\underline{a}$ falls under the very concept which defines $\underline{a}$; that is:

$\underline{a} \in \{x\colon s(x) \wedge (x \notin x)\}$,

or $\underline{a} \in \underline{a}$; but also $\underline{a} \notin \underline{a}$. Here the connection between vicious circularity and self-contradiction is at its clearest.

It therefore looks as though the (VCP) would eliminate all the antinomies by invalidating the arguments which generate them. Before examining this solution in any detail, we should discuss another methodological rule which was put forward by Russell and only half-heartedly accepted by Poincaré.

[D] Russell's Principle of Syntactical Prevention

This principle, which proves difficult to formulate in any rigorous way, enjoins the logician to pre-empt rather than to cure. Poincaré and Russell agreed about the following diagnosis: the paradoxes have to do, not with the *falsity* of assumptions like the Comprehension Axiom, but with some *logical nonsense* arising from the presence of vicious circularities in certain arguments. That is to say: the antinomies flow from a logical malformation. Russell criticized Poincaré for taking cognizance of vicious circularities only when it was too late to do anything about them. Poincaré's method in effect consists in starting on a road which is *subsequently* found to lead to an antinomy, hence in identifying vicious circularities *after* their having been committed, and in finally concluding, *post hoc,* that some arguments are globally inadmissible. In modern parlance: Poincaré makes his formation rules depend on the results of presumed proofs some of which are subsequently pronounced invalid. He writes down, in the object-language, sequences of expressions which may or may not be meaningful; by going over to the metalanguage, he decrees that certain sequences constitute no legitimate deductions and hence that some of their members are not well-formed. Russell criticizes Poincaré for not nipping the paradoxes in the bud—for not proposing a general principle which would, right from the start, guard us against falling into contradictions. Russell wants the antinomies to become unmentionable; and in his opinion, one does not make something unmentionable by first citing it and then maintaining that it cannot be mentioned. To put it a little differently, one does not circumvent a badly conducted argument by writing it down, even if only in order to declare it—*post hoc*—illegitimate. Despite being a fruitful heuristic device, the (VCP) does not, according to Russell, constitute an effective *remedy* against vicious circularity. Our logical grammar must be such as to render all paradoxes unthinkable or inexpressible. This is precisely why Russell opts for a *meaning* rather than for a *truth* or *provability* criterion. Whereas Zermelo *proves* $\neg(\exists b)[s(b) \wedge (\forall x)((x \in b) \leftrightarrow (x \notin x))]$, that is, he *refutes* the existence of any set of all sets, Russell regards both $(x \in x)$ and $(x \notin x)$ as meaningless and hence as devoid of truth-value. He consequently blamed Poincaré for sharing his viewpoint *without* trying to lay down a criterion which would ban all circular arguments as nonsensical. With regard to the antinomies, Poincaré was not really interested either in the meaningfulness or in the truth-values of any supposedly problematic axioms but in the possibility of performing certain mathematical operations. He tried to establish, by a *reductio*, that some purported constructions cannot be successfully carried out in time; which is why he often initiated certain operations in order to show either that they are constantly undone by subsequent constructions or that they cannot be brought to a close. The cornerstone of Poincaré's philosophy of mathematics thus seems to be, not the (VCP), but the following thesis, henceforth denoted by [P]:

> **[P]** Mathematical entities do not pre-exist to their definition, that is: to their construction in pure time. The intuition of subjective time is therefore needed in order to license the iteration of operations leading to the construction of mathematical objects. Hence no entity whose genesis is incompatible with the structure of temporal succession, for instance with the transitivity or anti-symmetry of the time-ordering relation, can possibly exist.

Principle [P] marks the parting of the ways between Poincaré and Russell, for the latter explicitly denied that temporal considerations had any role to play in the foundations of mathematics: he adopted a resolutely anti-Kantian stance. The following quotations demonstrate that Poincaré took a very different view.

> What is, then, the condition necessary for the rules of this logic to be valid? It is that the classification which is adopted be *immutable*. We learn that two soldiers are members of the same regiment, and we want to conclude that they are members of the same brigade; we have the right to do this provided that during the time spent carrying on our reasoning one of the two men has not been transferred from one regiment to another.

> The antinomies which have been revealed all arise from forgetting this very simple condition: a classification was relied on which was not immutable and which could not be so; the precaution was taken to *proclaim* it as immutable; but this precaution was insufficient. It was necessary to render it immutable in fact, and there are cases in which this is not possible. . . .
>
> But here again it is necessary that this law of correspondence be predicative. If we deal with two infinite collections, it will never be possible to imagine these two collections as exhausted. If we suppose that we have taken a certain number of objects in the first collection, the law of correspondence will enable us to define the corresponding objects of the second. If we then introduce new objects, this introduction of new objects might change the sense of the law of correspondence in such a manner that the object A' of the second collection, which before this introduction corresponded to an object A of the first collection, will no longer correspond to it after this introduction. In this case the law of correspondence will not be predicative.
>
> When I speak of all the integers, I mean all the integers which have been invented and all those which could be invented some day. When I speak of all the points in space, I mean all the points whose coordinates can be expressed by rational numbers or by algebraic numbers, or by integrals, or by any other method that can be invented. And it is this "that can" which is the infinity. (Poincaré 1913a, pp. 45, 49, 60)

In these passages, Poincaré deliberately chooses the temporal mode of speech in discussing the foundations of mathematics. It is therefore no accident that he saw in his controversy with the logicists and with the set-theorists a renewal of the old conflict between the Kantians and the Leibnizians.

In his seminal work on Poincaré's philosophy of mathematics, Gerhard Heinzmann regards 'predicativity' as the central tenet of Poincaré's pragmatism (Heinzmann 1985). Heinzmann rigorously formulates the successive versions of the (VCP) which were proposed by Poincaré between 1905 and 1912. He considers the last one—denoted by P_2—to be a new departure which Poincaré however failed to acknowledge as such (Heinzmann 1985, p. 68). Heinzmann compares and contrasts Poincaré's various definitions of predicativity in order to determine their comparative logical strengths; he tries, for example, to decide which is the most liberal and which the most restrictive among them. Since he proceeds by counter-examples, he could obtain only negative results of the form: P_{1a} is not wider than P_2; so he correctly concludes that "the chosen examples are not such as would dissipate all obscurity in the definition of the predicativity rules and in their application" (Heinzmann 1985, p. 76).

It is admittedly interesting and, from a logical viewpoint, even essential to compare and contrast different notions of vicious circularity. But given the multitude of definitions attributed to Poincaré, it seems obvious that we have to do with mere corollaries flowing from one fundamental proposition, namely from [P], which is precisely why Poincaré did not look upon P_2 as a really new criterion; for since P_2 follows from [P], he was taking cognizance of a principle to which he had in effect always implicitly subscribed.

By going back to [P], I propose first to explain why Poincaré rejected all notions of actual infinity, including that of the set ω of natural numbers; secondly, to throw light on his reasons for objecting to Russell's theory of ramified types; and then to analyse the sense in which [P] entails each of the (VCP)'s attributed by Heinzmann to Poincaré; finally, to account for the connection which Poincaré tried to establish between vicious circularity on the one hand, and actual infinities on the other.

To begin with, no Kantian like Poincaré could admit the existence of actual infinities, not even those of the denumerable kind. It was taken for granted that every mental operation had a duration greater than some strictly positive minimum. Though the latter might depend on the type of act to be performed, no construction of any infinite class can be completed in any finite time interval. This is a major reason for Poincaré's critical attitude towards Russellian logic, to which we now turn.

5.4 A Critique of Russellian Logic

[A] Russell's Latent Constructivism

In *Science and Method*, Poincaré discusses a number of conjectures envisaged by Russell as remedies for the paradoxes. The most important amongst them are: the Principle of Limitation of Size which we have already encountered, and the no-class theory which we are about to explain. Russell quickly abandoned the method of size-limitation as illustrated by Zermelo's definition of set-theory, for it appeared to him bound to issue in an adhoc solution. By appealing to Principle [A] (compare 5.3 above), Russell identified a trait common to all the paradoxes, namely a self-reproductive capacity which Jules Vuillemin formalized as follows:

(13) $(\forall u)[(s(u) \wedge (\forall x \in u)\psi(x)) \rightarrow (\exists f)(\psi(f\text{'}u) \wedge (f\text{'}u \notin u))] \rightarrow$
$\neg(\exists u)[s(u) \wedge (\forall x)(x \in u) \leftrightarrow \psi(x)]$

In other words: let $\psi(x)$ be any predicate having the following property: for any set u all of whose members fall under ψ, there exists a function f such that the image f'u of u under f both satisfies ψ and falls outside u; then the extension of ψ cannot be a set. It is well-known that this notion of a self-reproductive process led Russell to his theory of logical types.

Turning to the the no-class thesis, let us first note that Poincaré does not conceal his aversion to it; which, given his adherence to Aristotelian Logic, is hardly surprising:

> In the no-class theory all mention of the word class is prohibited, and the word has to be replaced by various periphrases. What a change for the logisticians who speak of nothing but class and classes of classes! The whole of Logistic will have to be refashioned. Can we imagine the appearance of a page of Logistic when all propositions dealing with class have been suppressed! (Poincaré 1908, p. 188)

According to Heinzmann, Russell did not achieve his logical objectives by means of the no-class theory which was 'never presented' in any detail and was subsequently replaced by the theory of types. Once again, I beg to differ both from Poincaré and from Heinzmann. To begin with: the no-class thesis forms an integral part of *Principia Mathematica*, Whitehead's and Russell's definitive work on logic. It will moreover be shown that type theory and the elimination of classes are two complementary aspects of a unified solution to the paradoxes. And it finally turns out that in accordance with Principle [C] of 5.3 above, both of these aspects are supported by a logical intuition totally independent of the antinomies.

As already explained, it appeared to Russell that any attempt merely to reduce the scope of the Comprehension Scheme would spawn adhoc hypotheses and thus violate Principle [C]. It remained to be seen whether the category of classes, which is central to the Comprehension Axiom, can be *ontologically* dispensed with. Note that the reference to ontology is essential, for we still want to be able to use 'class' as shorthand for something more fundamental. This seems *prima facie* possible. When we assert that all men are mortal, we do not normally start by constituting the set of all men, then that of all mortals, in order finally to assert that the first class is contained in the second; we directly perceive a relationship between two predicates, namely those of humanity and of mortality. At least in this case, our intuition supports the view that classes are dispensable fictions. Meyerson pointed out that the human mind finds it easier to reason about objects than about properties; whence its tendency to hypostasize all predicates into classes. But precisely because we are dealing here with a pronounced *tendency* of our minds, Meyerson further maintained that we *impose our reifications* on nature (Meyerson 1921, Chapter 5). This Kantian aspect of Meyerson's philosophy is therefore compatible with the non-existence of classes. As for Russell's thesis, it tells us that for any propositional formula **K**(a), â**K**(a) should be treated as an incomplete class-symbol contextually defined by the following schemes:

(14) $\mathbf{B}(\hat{a}\mathbf{K}(a)) \equiv_{\text{Def.}} (\exists\varphi)[(\forall x)(\mathbf{K}(x) \leftrightarrow (x \in \varphi)) \wedge \mathbf{B}(\varphi)]$;
where $(x \in \varphi) \equiv_{\text{Def.}} (\varphi!(x))$ and $\varphi!$ indicates that φ is a predicative variable, i.e one whose order is the immediate successor of the maximum of that of its arguments.

Neither ∈ nor ! belongs to the primitive vocabulary of Russell's system (see below, pp. 000–000).

Russell's solution thus reconciles two conflicting tendencies of the human mind: on the one hand, its proclivity towards hypostasizing certain processes in order more conveniently to think about them as *things*, and on the other, its propensity to deny the real existence of classes. Without fear of contradiction, we can normally proceed as if classes really existed; but in situations which seemingly lead to contradictions, a translation based on (14) reveals a category mistake which might initially have escaped detection. For example, when translated into the primary language of the *Principia*, Russell's antinomy reduces to the pseudo-formula:

(15) $(\exists\psi)(\forall\varphi)[\psi(\varphi) \leftrightarrow \neg\varphi(\varphi)]$.

This string of symbols is instinctively felt to be meaningless, or non-well-formed, for we all share Frege's intuition that in contradistinction to objects, all predicates are stratified. Frege started from a fundamental demarcation between function and object: the former, but not the latter, is saturated. These two entities thus lie in disjoint categories (see Frege 1986). Note also that according to Frege, a proposition denotes a truth-value and hence an object. It follows that all predicates occurring in a sentence must ultimately be completed either by individual variables or by proper names; for no entity becomes saturated through being joined to another unsaturated entity. It would be absurd to ask whether or not (the property of) being red is in turn red: 'is red is red' is a meaningless phrase since the first 'is red' still needs completion. For the same reason, both $\varphi(\varphi)$ and $\neg\varphi(\varphi)$ lack sense, thus making Russell's paradox inexpressible. Starting from his fundamental distinction between object and function, Frege proceeds to a general differentiation—by type—of properties and relations: a property is said to be *first-order* if and only if (saturated) objects can conceivably fall under it; a relation some of whose argument-places are filled by first-order predicates must therefore basically differ from the latter through possessing a different type of argument, namely an unsaturated predicate rather than a proper name; and so on. Let us give a few examples. A relation between objects can fall under the higher-order property of transitivity. Thus transitivity subsumes relations which apply, in turn, to individual objects. Another illustration is provided by the concept of existence which Frege regards as a property, not of individuals, but of predicates. He takes 'P exists' to have the same meaning as $(\exists x)P(x)$, i.e. as $[\lambda Q(\exists x)Q(x)](P)$. $[\lambda Q(\exists x)Q(x)]$ obviously expresses a second-order property under which the predicate P might fall.

Going back to Russell: by means of the no-class theory, he first rid himself of all sets while keeping their intensions intact; he then dissolved the paradoxes by introducing type-distinctions between various predicates. We have here two prongs of one global solution; and Russell rightly claimed that his type-theoretic approach was consonant with logical commonsense. Since he kept at his disposal all the predicates he started from, the expressive power of his system remained unimpaired, thus enabling him to reconstruct the whole of known mathematics. There was however no guarantee that all the postulates he needed would have a strictly logical character. Moreover: the solution just outlined—which came to be known as the 'simple theory of types'—eliminates the set-theoretical contradictions but not the vicious circles responsible for the linguistic paradoxes. Thus both from Poincaré's and from Russell's fundamentalist viewpoints, simple types could yield no satisfactory solution to all the antinomies.

It should be noted that right from the beginning, Russell envisaged the theory of *ramified* types which incorporates the (VCP) and constitutes the radical answer required by Principle [A] above. Without being fully aware of it, Russell simultaneously followed two different guidelines. The first led him to Frege's conception of *simple types*, and the second to that of the *order* of a type and of every well-formed expression. This second notion demands that whenever a statement refers to some totality, its order be raised by at least one unit above that of any member of the totality in question. The differentiation by order thus represents Russell's version of the (VCP). The fusion of these two intuitions which, from the very beginning, was instinctively carried out by Russell gives rise to the ramified theory of types. In the absence of something like the constructivism professed by Poincaré, it is however impossible to justify the hierarchy of ordered types in any non-adhoc manner, that is: independently of the semantic antinomies. To the extent that he was a realist and even a Platonist, Russell had therefore ended up proposing an adhoc solution to the linguistic paradoxes. The following passage shows that in any case he resorted to constructivist modes of reasoning whenever he tried to defend his own VCP:

> That is to say, a function is not a well-defined function unless all its values are *already* well-defined. It follows from this that no function can have among its values anything which presupposes the function, for if it had, we could not regard the objects ambiguously denoted by the function as definite *until* the function was definite, while conversely, as we have just seen, the function cannot be definite *until* its values are definite. This is a particular case, but perhaps the most fundamental case, of the vicious circle principle. (Russell 1973, pp. 217–18. My italics).

Let us note the occurrence, in this passage, of words like 'already' and 'until' whose temporal import cannot be doubted. Without admitting it, Russell had thus adopted a viewpoint closely resembling Poincaré's. Of the two thinkers, Russell was admittedly the one who technically succeeded in recasting the foundations of mathematics in accordance with the (VCP). He did not however acknowledge that in as far as it contains a (VCP) based on the notion of order, his theory of ramified types presupposes constructivism as its only philosophical foundation. By contrast, Poincaré conceded that vicious circularities could not be condemned from the realist standpoint which he nonetheless rejected. The following quotation shows that he deliberately founded his mathematical 'pragmatism' on temporal considerations:

> But there is a third type of definition, and this is the origin of a new misunderstanding between the pragmatists and the Cantorians. These are again definitions by postulates, but the postulate here is a relation between the object to be defined and *all* the individuals of a genus of which the object to be defined is supposed to be a member (or of which one supposes to be members objects which themselves can be defined only by the object to be defined). This is what happens if we posit the following two postulates:
>
> *X* (object to be defined) is related in such and such a way to *all* members of the genus *G*.
>
> *X* is a member of *G*.
>
> Or else the following three postulates:
>
> *X* is related in such and such a way to *all* the members of the genus *G*.
>
> *Y* is related insuch and such a way to *X*.
>
> *Y* is a member of *G*.

> To the pragmatists such a definition implies a vicious circle. It is not possible to define *X* without knowing all the members of the genus *G*, and consequently without *X* which is one of these members. The Cantorians do not admit this; the genus *G* is given, consequently we know *all* its members. The definition has as purpose only to *distinguish* from among these members the one which has the expressed relation with all its fellow-members.
>
> 'No', reply their opponents, 'the knowledge of the genus does not result in your knowing all its members; it merely provides you with the possibility of constructing them all, or rather of constructing as many of them as you wish. They will exist only after they have been constructed; that is, after they have been defined; *X* exists only by virtue of its definition, which has meaning only if all the members of *G* are known beforehand, and *X* in particular. (Poincaré 1913a, pp. 70–71)

This passage gives a complete account of Poincaré's philosophy of mathematics. Gödel came later to share Poincaré's point of view on one vital issue: for a believer in a pre-existent mathematical reality, it is by no means absurd to describe an object by reference to a totality containing that self-same object; for the totality and hence the object itself are supposed to be given prior to our thinking about them; or more precisely: they are taken to dwell in an atemporal Platonic universe (Gödel 1983b). It is no accident that Gödel discovered his constructibility axiom V = L through an analogy with Russell's theory of types: in the constructible universe L, the ordinal μ represents the order of the stratum L_μ. It is also hardly surprising that qua Platonist, Gödel regarded V = L as false.

[B] Wittgenstein's Paradox and the Primacy of Arithmetic

Wittgenstein claimed to have discovered a paradox in Russell's *Principia*. After asserting the necessity of assigning a type to *every* propositional function, Russell went on to maintain that quantifications ought to be restricted to one type at a time. He thus appeared to violate his own injunctions by referring to *the totality* of propositional functions. It seemed as though whenever he tried to describe his system, he could not avoid transgressing its rules. Wittgenstein concluded that philosophy offers neither genuine arguments nor any genuine propositions bearing a definite truth-value. This is Wittgenstein's well-known *ineffability thesis:* philosophy is required neither to found nor even to explain a system on which all reasoning is meant to be based; it must content itself with *exhibiting* certain relations and with *displaying* rather than with establishing logical facts. Russell could have replied that in describing his own system, he did not have to refer to any propositional *functions* as such but only to certain syntactically well-formed strings of symbols; these are all of the same type and can therefore be legitimately quantified over. Of course, we intuitively know that these inscriptions are intended to denote functions of various types; but in presenting his grammar, Russell did not have explicitly to mention this semantic aspect of his system. In other words, Russell could have adopted Hilbert's formalist attitude by giving a purely syntactical description of the theory of ramified types. So Wittgenstein's ineffability thesis need never have been taken seriously. There remains however one major difficulty which Poincaré had—long before Wittgenstein—clearly pointed out: even though we might restrict ourselves to the syntactical aspect of a formal system, we still have to lay down rules for the combination of its primitive symbols; and for these operations to make any sense, arithmetic had to be presupposed. He wrote:

> 1. In this hierarchy there occur without difficulty propositions of the first, second, order etc., and in general of the nth order, n being any finite integer. Is it possible to consider similarly propositions of order α, α being a transfinite ordinal number? It is thus that Mr König thought of a theory which does not differ essentially from that of Mr Russell. . . .
> 2. If we answer 'yes' to the first question, it will be necessary to explain what is understood by the objects of order ω, ω being ordinary infinity, that is the first transfinite ordinal number; . . .
> 3. If, on the other hand, we reply 'no' to the first question, how shall it be possible to base on the theory of types the distinction between finite and infinite numbers, since this theory is devoid of meaning if it is not assumed that this distinction has already been made?
> 4. More generally, whether we reply 'yes' or 'no' to the first question, the theory of types is incomprehensible, if we do not suppose the theory of ordinal numbers already established. How will it then be possible to base the theory of ordinal numbers on that of types? (Poincaré 1913a, p. 52)

Russell answered Poincaré's criticism in 'The Theory of Logical Types', which was published—in French—in the 1910 issue of the *Revue de Métaphysique et de Morale*:

> This assertion appears to me to rest on a confusion. That the types have an order is admitted; but it is not admitted that it is necessary to study this order as an order. The steps of a deduction have an order, but it is not necessary to the deduction to study the order of the steps, although, when we turn our attention to the order, we see that it is essential to the deduction. So with types: they have an order, and when we study it, we see that it is important. But we can make all the uses of them that are required without studying the order, just as we can distinguish a function $\varphi\hat{a}$

> from a function φ(â,û) without knowing that the first has one argument while the second has two, though it would be pedantry to avoid all phrases which involve this knowledge, as soon as we have seen that we could avoid such phrases if we chose. So, with types, we may speak of their order in words which, strictly speaking, involve a knowledge of the ordinals, because it is obvious that we could make all the necessary uses of types without such words. Instead of speaking of functions of the first order, we could speak of 'functions φ!â'; instead of functions of the second order, functions f!(φ!â), and so on. Thus, although types have an order, the ordinals are not pre-supposed in the theory of types, and there is no logical circle in subsequently basing the theory of ordinals upon a basis which assumes the theory of types. (Russell 1973, p. 252)

It can be maintained both that Russell understood perfectly well the difficulty raised by his critic and that he was right but misleading in his reply to Poincaré's criticisms. Russell's position is genuinely *para-doxical* in that it displays perplexing aspects, which do not however add up to a *formal* contradiction. We have already referred to recursive enumerability as to the essential characteristic of any axiomatisable system S. It is therefore possible to stumble on a device which would mechanically churn out a potentially infinite proof, D say, of all the theorems of S. To claim, in these circumstances, that D 'presupposes arithmetic' would certainly make no sense; but an *understanding* even of the bare syntactical structure of D must be by way of induction, which presupposes a large part of arithmetic. It has been maintained that we do not have here a fact of logic but of human psychology. This claim is however mistaken: no matter how large the number n, neither the first n members nor even the whole of the sequence D can capture the *objective* structure intended by D. This is why the grammar of S has to be written down; and for this we need an informal, which is to say: an *already interpreted* arithmetic. We would for example have to show that the property of being an axiom is decidable; which in turn presupposes the unique determination of the components into which any formula splits; and this uniqueness is established by induction over the degree of logical complexity of a wff: one has to decide whether or not a given wff is of the form $[(\forall x)\mathbf{K}(x) \rightarrow \mathbf{K}(t)]$, where t is a term having the same type as the varible x; and one ought to have previously established that if $\mathbf{K}(x)$ is well-formed, then so is $\mathbf{K}(t)$. This is again proved by induction on the degree of complexity of $\mathbf{K}(x)$. The rules of inference raise similar difficulties: in order to determine whether or not A,B $\vdash$ C is an instance of Modus Ponens, one should be able to decide recursively whether $B \equiv (A \rightarrow C)$. Poincaré was therefore right to insist on the indispensability of mathematical induction. And yet Russell was not altogether wrong in pointing out that a description of the grammar of the system S forms no genuine part of S. (Let us note that we have here the origins both of Wittgenstein's ineffability thesis and of Russell's distinction between object-language and meta-language).

In 1927 Bernays arrived at the same conclusion as Poincaré. Henceforth, the primacy of logic over mathematics could no longer be maintained: at a presystematic level, we may possess a logical intuition entirely independent of mathematics; but in setting up any formal logical structure, we have to make use of arithmetic, so that even if a theory purported to provide a foundation for the whole of mathematics, its syntax would already exhibit an arithmetical and hence a mathematical structure. There can be no question either of the primacy of logic over mathematics or of the converse: the two disciplines have to be simultaneously deployed (see Bernays 1976b).

Gödel said of Russell's *Principia* that from the viewpoint of syntactal rigour, they represented regress as compared with Frege's achievement. And we cannot help being surprised at finding, on the one hand that Russell hardly makes any mistakes in handling complex formal expressions; on the other, that he never rigorously defines the grammar of his own system. For example, in order to convey what he means by the type of a propositional function, he proceeds by a series of examples followed by an 'etcetera'. We are still far from the modern, rigorous presentations of formal languages; which is why the *Principia* have often been described as constituting a *semi*-formal system. Having closely followed Frege's work, Russell must nonetheless have known that no 'etcetera' should occur in any formal definition; and his contemporary and friend, the Cambridge mathematician G.H. Hardy, taught analysis from a Weierstrassian viewpoint which was based on clear definitions of the concepts of limit, of continuity and of convergence. Russell himself had rigorously characterized the real numbers as transitive classes of rationals possessing no maximum element, so he was perfectly capable of describing his syntax in a more satisfactory way. We can therefore be rightly astonished at the disparity between the rigorous character of his system on the one hand, and his somewhat sloppy way of presenting it on the other. He could, after all, have started the *Principia* by giving the following recursive definition of a ramified type t and of its order $\mu(t)$:

(15) For every natural number n, the numeral $\underline{n}$ is a type, and $\mu(\underline{n}) = n$.

(16) Given any finite sequence of types $t_1, t_2, \ldots, t_n$ and any natural number m such that $m > \mu(t_i)$ for all $i = 1,2,..,n$, then $\underline{m}[t_1,t_2,\ldots,t_n]$ is also a type and: $\mu(\underline{m}[t_1,t_2,\ldots,t_n]) = m$.

(17) All ramified types are obtained through a finite number of successive applications of (15) and (16). (For the sake of simplicity, m will often be written for $\underline{m}$; i.e. we shall not distinguish between a numeral and the corresponding natural number.)

Russell could then have introduced his basic vocabulary as follows: for each type t, there would be two denumerably infinite sequences of constants: a_{tn}, and of variables x_{tn} $(n = 0,1,2,...)$; where t is the type both of a_{tn} and of x_{tn}, while n is a mere label. To these symbols would be added the following logical connectives: $\neg$, $\vee$, $\exists$, $\wedge$ (where $\wedge$ is in effect Church's λ-symbol. In what follows we shall often use λ in preference to $\wedge$). There would also be the two brackets,) and (. All well-formed expression (wfes) **K** and their type $\underline{\mu}(\mathbf{K})$ would be recursively defined as follows:

(18) Both a_{tn} and x_{tn} are wefes, and $\underline{\mu}(a_{tn}) = \underline{\mu}(x_{tn}) = t$.

(19) Let $\mathbf{K},\mathbf{B_1},..., \mathbf{B_n}$ be wefes of types $m[t_1,...,t_n],t_1,..., t_n$ respectively; i.e. let $\underline{\mu}(\mathbf{K}) = m[t_1,...,t_n]$ and $\underline{\mu}(\mathbf{B_j}) = t_j$ for all $j = 1,...,n$; where m is a natural number. (In view of (16) above, $m > \mu(t_1)$, so that $m \geq 1$). Then $(\mathbf{K}(\mathbf{B_1},...\mathbf{B_n}))$ is also a wefe and $\underline{\mu}(\mathbf{K}(\mathbf{B_1},...,\mathbf{B_n})) = m$.

(20) Let **K** be any wefe such that $\underline{\mu}(\mathbf{K}) = m$, where m is a nonzero natural number; and let $z_1,...,z_n$ be distinct variables of types $t_1,...,t_n$ respectively. Then: $((\lambda z_1...z_n)\mathbf{K})$ is a wefe and $\underline{\mu}(((\lambda z_1...z_n)\mathbf{K})) = k[t_2,...,t_n]$, where $k =$ $\text{Max}(m, \mu(t_1) + 1,...,\mu(t_n) + 1) =$ $\text{Max}(\underline{\mu}(\mathbf{K}), \mu(\underline{\mu}(z_1)) + 1,\ldots,\mu(\underline{\mu}(z_n)) + 1)$.

(21) Let K_1 and K_2 be wefes such that both $\underline{\mu}(K_1)$ and $\underline{\mu}(K_2)$ are positive integers. Then $\neg(K_1)$ and $(K_1 \vee K_2)$ are wefes, and: $\underline{\mu}((\neg K_1)) = \underline{\mu}(K_1)$, $\underline{\mu}((K_1 \vee K_2)) = \text{Max}(\underline{\mu}(K_1),\underline{\mu}(K_2))$. Thus both $\underline{\mu}((\neg K_1))$ and $\underline{\mu}((K_1 \vee K_2))$ are nonzero natural numbers.

(22) Let **K** be any wefe such that $\underline{\mu}(\mathbf{K})$ is a positive integer. Then $(\exists x_{tn}\mathbf{K})$ is also a wefe, and $\underline{\mu}((\exists x_{tn}\mathbf{K})) = \text{Max}(\mu(t) + 1, \underline{\mu}(\mathbf{K}))$; so that $\underline{\mu}((\exists x_{tn}\mathbf{K}))$ is also a positive integer.

A wefe **K** is said to be a well-formed formula, or a wff, iff its type $\underline{\mu}(\mathbf{K})$ is a nonzero natural number. In such a case, **K** represents a true-or-false proposition, and the type $\underline{\mu}(\mathbf{K})$ of **K** is at the same time the order of **K**, or more pedantically: the order of the type of **K**. Rule (22) constitutes Russell's (VCP): the order of $(\exists x_{tn}\mathbf{K})$, where x_{tn} is the variable over which the quantification is carried out, is greater by at least one unit than that of x_{tn}.

The grammatical definitions given above could easily have been proposed in 1908. Such a move would however have exposed Russell to Poincaré's objection that while purporting to found the whole of mathematics, the new Logistic actually presupposes induction and hence arithmetic. Russell's use of examples expressed in ordinary English—and ending in 'and so on'—conceals the necessity of invoking arithmetic in order to describe the structure of the *Principia.* The only way of presenting an all-embracing system, one which would serve as a basis for all deductive thought, is in fact the method which a Cambridge logician threatened to use in his lectures. He proposed to give his course by first writing the sequence of all the primitive symbols of his system; by then exhibiting all the wffs involved in the proofs of certain theorems; by finally writing down the proofs, while avoiding—throughout—pronouncing a single word of any natural language. He hoped that by some sort of inductive process, the students would understand what was meant by a well-formed expression, by an axiom and by a rule of inference. Mercifully, he never fully carried out his programme, but he came pretty close to doing so. Those who regarded him as an eccentric were however mistaken: he simply took very seriously Wittgenstein's view that the foundations of logic could only be *shown* but not *explained* in any meaningful way.

Let us go back to Russell who refused to accept Wittgenstein's ineffability thesis. This ultimately led him to propose the distinction between object-language and meta-language which Tarski later used in establishing his famous result; namely that the class of theorems of the most important formal theories is strictly contained in that of all true statements. In the 1922 preface to the *Tractatus*, Russell describes his uneasiness about Wittgenstein's position as follows:

> These difficulties suggest to my mind some such possibility as this: every language has, as Mr Wittgenstein says, a structure concerning which, in the language, nothing can be said, but that there may be another language dealing with the structure of the first language, and having itself a new stucture, and that to this hierarchy there may be no limit. Mr Wittgenstein would of course reply that this whole theory is applicable unchanged to the totality of such languages. The only retort is to deny that there is any such totality. The totalities concerning which Mr Wittgenstein holds that it is impossible to speak logically are nevertheless thought by him to exist, and are the subject-matter of his mysticism. The totality resulting from our hierarchy would be not merely logically inexpressible, but a fiction, a mere delusion, and in this way the supposed sphere of the mystical would be abolished. (Wittgenstein 1972, Preface)

Once again, Russell's remarkable logical intuition leads him to introduce a stratification into object and meta-level which parallels his earlier distinction between different types within the same language. It should however be noted that he infringes his own rules by holding that *every* system could give rise to a new metalanguage; he thus quantifies over the class of *all* languages and hence over a totality declared by him to be illegitimate. One way of avoiding both this paradox and the infinite regress from *every* language to *its* metalanguage consists in taking, at some finite stage, an informal 'meta-arithmetic' as simply given. Such an arithmetic would be intuitively interpreted: it would be *inhaltlich* in Hilbert's sense. It could be founded on the pure intuition of time which, according to both Kant and Poincaré, enables us to generate the number sequence; it would moreover provide us with the inductive principle needed for describing the structure of the most fundamental formal system, namely that of deductive logic. In this way, Poincaré's views would have been vindicated, if only posthumously.

[C] The Partial Failure of Logicism

The logicist programme was two-pronged. First, all mathematical concepts were to be defined in purely logical terms, a task which was successfully carried out by Russell. The known theorems had secondly to be derived from strictly logical axioms. During this second stage, Russell became aware of the indispensability of postulates whose logical status was questionable. These were the axioms of Choice, of Reducibility, and of Infinity. In the language of ramified types, the Axiom of Choice takes the form:

(23) $(\forall R)(\exists S)[(\forall x,y,z)((S(x,y) \wedge S(x,z)) \rightarrow (y = z)) \wedge (\forall x,y)(S(x,y) \rightarrow R(x,y)) \wedge (\forall x)((\exists y)R(x,y) \rightarrow (\exists y)S(x,y))]$,

where R and S have the same type, $m[t_1,t_2]$ say; and x, y, z are of types t_1, t_2, t_2 respectively. Thus (23) asserts that every relation R contains a function S whose domain coincides with that of R.

The Axiom of Reducibility actually consists in the following *scheme*:

(24) $(\exists P)(\forall x_1,x_2,\ldots,x_n)[P!(x_1,x_2,\ldots,x_n) \leftrightarrow \mathbf{K}(x_1,x_2,\ldots,x_n)]$; where:

$\mathbf{K}(x_1,\ldots,x_n)$ is an arbitrary open wff in which $x_1,\ldots, x_n$ may occur free; P is a variable which does not occur free in $\mathbf{K}(x_1,\ldots,x_n)$; and, as already explained, the exclamation mark in $P!(x_1,\ldots,x_n)$ indicates that P is *predicative*, i.e. that P has type $p[t_1,\ldots,t_n]$, where: $t_1,\ldots, t_n$ are the types of $x_1,\ldots, x_n$ respectively, and $p = Max(\mu(t_1),\ldots,\mu(t_n)) + 1$. p is thus the smallest natural number such that $p[t_1,...,t_n]$ is an allowable (ramified) type. (Note again that the sign ! does not belong to the primitive vocabulary of the object language.)

For the sake of simplicity, we shall express the Axiom of Infinity in the language of simple types:

(25) $(\forall m)[N(m) \rightarrow (\exists P)m(P)]$;

where N(m) and m(P) respectively mean that m is a natural number and that exactly m elements fall under P. Thus Axiom (25) tells that, for every m, there exist m distinct objects; from which it clearly follows that the universe cannot be finite.

Let us first examine the Axiom of Reducibility, which has given rise to notorious controversies between warring schools of thought. Poincaré initially took scheme (24) to be equivalent to the Principle of Induction. Though made aware of his mistake by Russell, he still inclined towards regarding reducibility and induction as twin *synthetic apriori truths.* Paradoxically, it was the *logicist* Russell who was prepared to go much further by holding Reducibility and Choice to be contingent assumptions. Convinced of the latter's non-obvious character, Russell ended up claiming that similarly to astronomy, logic was a hypothetico-deductive science whose truths were *synthetic a posteriori*:

> The method of logistic is fundamentally the same as that of every other science. There is the same fallibility, the same uncertainty, the same mixture of induction and deduction, and the same necessity of appealing, in confirmation of principles, to the diffused agreement of calculated results with observation. The object is not to banish 'intuition', but to test and systematise its employment, to eliminate the errors to which its ungoverned use gives rise, and to discover general laws from which, by deduction, we can obtain true results never contradicted, and in crucial instances confirmed, by intuition. In all this logistic is exactly on a level with (say) astronomy, except that, in astronomy, verification is effected not by intuition but by the senses. The 'primitive propositions' with which the deductions of logistic begin should, if possible, be evident to intuition; but that is not indispensable, nor is it, in any case, the whole reason for their acceptance. This reason is inductive, namely that, among their known consequences (including themselves), many appear to intuition to be true, none appear to intuition to be false, and those that appear to intuition to be true are not, so far as can be seen, deducible from any system of indemonstrable propositions inconsistent with the system in question.
>
> Among several systems fulfilling all these conditions that one is to be preferred, aesthetically, in which the primitive propositions are fewest and most general; exactly as the law of gravitation is to be preferred to Kepler's three laws as the starting point of mathematical deductions. If intuition were infallible, this complicated

> process of verification would be unnecessary. But, as is shown by the contradictions, intuition is not infallible. Hence an element of uncertainty must always remain, just as it remains in astronomy. (Russell 1973, p. 194)

Russell held that any reason adduced in support of the Axiom of Reducibility would have to be strictly inductive. This is shown by the following quotations, also taken from *Essays in Analysis*:

> In the case of the Axiom of Reducibility, the inductive evidence in its favour is very strong, since the reasonings which it permits and the results to which it leads are all such as appear valid. But although it seems very improbable that it should be found to be false, it is by no means improbable that it should be found to be deducible from some other more fundamental and more evident axiom. It is possible that the use of the vicious-circle principle, as embodied in the above hierarchy of types, is more drastic than it need be, and that by a less drastic use the necessity of the axiom might be avoided. . . .
>
> Infallibility is never attainable and therefore some element of doubt should always attach to every axiom and to all of its consequences. In formal logic, the element of doubt is less than in most sciences, but it is not absent, as appears from the fact that the paradoxes followed from premises which were not previously known to require limitations. (Russell 1973, p. 251)

These passages express the whole of Russell's—clearly inductive—methodological creed. But given this standpoint, the word 'logicism' becomes an abuse of language whose continued use borders on intellectual dishonesty; for the concept of *logic* evokes that of truth in all possible worlds, a notion which applies to no synthetic hypothetico-deductive system; for the latter can in principle be *falsified* by the refutation of any one of its consequences, be this refutation intuitive or experimental.

Poincaré rightly maintained that for a logicist to be involved in contradictions is tantamount to receiving a death sentence. The notion of an inductive or of a hypothetico-deductive *logical* theory moreover constitutes a category mistake; for as just explained, such a system is judged by its consequences. Suppose that one of these is falsified. We would naturally have applied some logical rules in drawing such a consequence. Should we come to regard the underlying logic as having been undermined, then this might well mean giving up one or more of these rules. It is well-known that Quine reformulated Logic in such a way that only Modus Ponens need be used. Thus giving up even a single rule of inference might make all deductions impossible. We have furthermore to rely on logic in identifying the *non-logical* axioms in order then to conclude that at least one of them must have been refuted. But in the present case, we are dealing with a purely logical system, so there can be no question of there being any *non-logical* postulates. We therefore confront a situation very different from one involving an empirical science like astronomy. In Logic, we can rely only on our intuition to tell us which of our axioms are at fault and how they are to be modified; which is precisely the point made by Poincaré. In other words, it is only against some provisionally fixed logical background that hypothetico-deductivism works. Questioning logic would cast doubt on the hypothetico-deductive process itself; so the latter cannot be turned against its own logical foundations. Logic can certainly be criticized and improved but not by way of its 'contingent' consequences, whatever such a word might mean in this context. (We naturally except the case where one of these consequences is self-contradictory and hence non-contingent.)

Russell had thus manoeuvered himself into an impossible situation. He could of course have given up logicism and presented the *Principia* as a description of a possible Platonic universe. Like Frege, he had once been a Platonist and was not therefore committed to regarding logic as an empty system applying indifferently to all worlds; but Poincaré had in effect offered him a better alternative, a kind of honourable retreat. As mentioned above, Poincaré looked upon certain axioms like those of Choice and of Infinity as synthetic apriori propositions; and he had been essentially right in that the Axiom of Choice is a quasi-logical principle based on the use of the indefinite article in many natural languages. As for the Axiom of Reducibility, we shall now show that Russell himself ought to have had good reasons for holding it to be a quasi-logical assumption.

Note first that the Axiom of Reducibility does not, as has sometimes been claimed, reintroduce any vicious circularities into mathematics. In (24), the order of P can of course equal that of a bound variable occurring in $\mathbf{K}(x_1,...,x_n)$. But (24) does not generally entail $(\exists P)[((\lambda x_1...x_n)\mathbf{K}(x_1,...,x_n)) = P!]$. In most cases, this last expression will in fact be meaningless; for the type of $((\lambda x_1...x_n)\mathbf{K}(x_1,...,x_n))$ is, by definition, equal to $m[t_1,...t_n]$ where: q, $t_1,..., t_n$ being the types of $\mathbf{K}(x_1,...,x_n)$, $x_1,..., x_n$ respectively, we have $m = \mathrm{Max}(q, \mu(t_1) + 1,..., \mu(t_n) + 1)$. It is thus possible for $m = q > \mu(t_i) + 1$ to hold for all $i = 1,2,...,n$; so that $m > p$, where $p = \mathrm{Max}(\mu(t_1),...,\mu(t_n)) + 1$. It would follow that $m[t_1,...t_n] \neq p[t_1,...,t_n]$ = type of P!. The string of symbols $[((\lambda x_1...x_n)\mathbf{K}(x_1,,x_n)) = P]$ would then be devoid of meaning since the two sides of the equation belong to different types. Though co-extensional with $((\lambda x_1...x_n)\mathbf{K}(x_1,...,x_n))$, P is not *defined* in terms of classes whose elements have the same type as P. Hence the Axiom of Reducibility does not violate the (VCP).

Apart from its compatibility with the (VCP), the Axiom of Reducibility springs from the same intuition which supports the Comprehension Scheme; except that this intuition is now corrected by an additional insight into the differences between various logical types. More precisely: although the no-class theory eliminates all extensions conceived as *things,* we still feel that two propositional functions $\mathbf{K}(x_1,...,x_n)$ and $\mathbf{E}(x_1,...,x_n)$ which are satisfied by the same n-tuples $<x_1,...,x_n>$ have *something* in common; where this 'something' is not a simple *object* but some higher-order entity, namely some predicative P!. We recall that the paradoxes arise not from the mere *existence* of certain classes, but from assuming the latter to be of the same type as their members. Yet we instinctively feel that a group of soldiers, an army, ought to be more abstract an entity than any of its elements. Going back to **K** and **E**: as long as we consider only the n-tuples $<x_1,...,x_n>$ falling under these two predicates, we require the type of their extension to be determined exclusively by those of $x_1,...,x_n$ and not by the logical structure, or by the Fregean sense of **K** and **E**. The variable P occurring in (24) satisfies this requirement since its type, namely $p[t_1,...,t_n]$, is uniquely determined by the equation: $p = Max(\mu(t_1),...,\mu(t_n)) + 1$, where $t_1,...,t_n$ denote the types of $x_1,...,x_n$ respectively. As already mentioned, p is the smallest natural number making $p[t_1,...,t_n]$ into an admissible type. The Axiom of Reducibility tells us that one predicate P norms all the propositional functions extensionally equivalent to $\mathbf{K}(x_1,...,x_n)$. We therefore have a quasi-logical axiom which replaces the Comprehension Scheme, avoids all vicious circularities and leaves Russell's no-class theory intact. Taking n = 1, we obtain from (24) the following special case of the Reducibility Scheme:

(24) $(\exists P)(\forall x)[P!(x) \leftrightarrow \mathbf{K}(x)]$.

Were we to write $(x \in P)$ for P(x), then (26) would be hardly discernible from the Comprehension Scheme. The only difference is that unlike b, P has an order higher than that of x. And one need only compare (26) with Zermelo's disparate special cases of the Comprehension Scheme, namely with the Pairing, Replacement, Separation, Union, and Power Set Axioms, in order to acknowledge the methodological superiority of Russell's approach over Zermelo's; all this despite Russell's dissatisfaction with Scheme (26). Had he adopted Poincaré's position according to which Reducibility is a synthetic a priori principle, then Russell would have given his *Principia* a firm *philosophical* grounding. It ought nonetheless to be added that by contrast with Reducibility, the Axiom of Infinity has a very doubtful logical status; so that the logicist programme had other formidable problems to contend with.

5.5 Poincaré's Partial Solution to the Paradoxes

Poincaré's solution remained incomplete in that it eliminated the contradictions without enabling its author to reconstruct the whole of mathematics in accordance with Principle [P] (see p. 000). As remarked above, it was [P] and not the (VCP) which constituted Poincaré's fundamental criterion for the existence of mathematical objects. In the light of [P], we have seen why he could not accept any actual infinities, no matter how small their presumed cardinalities might be. We shall now examine the sense in which [P] entails the different versions of the (VCP) which were attributed by Heinzmann to Poincaré; also the reason why Poincaré took vicious circularity to be closely connected with the alleged existence of actual infinities.

The word 'definition' will henceforth denote a sequence of operations intended to *generate* some mathematical entity. Every other kind of definition—which might single out an already existing object—will be called a 'determination'. Poincaré held *all formally correct* determinations, including those involving circularities, to be legitimate. Hence a *determination* which circumscribes an object by reference to a totality some of whose members may depend on the object in question is permissible; for by using the word 'determination', we implicitly indicate that all the elements of the totality have already been constructed; so that as required by [P], the laws of temporal succession will have been obeyed. In other words, once certain entities have been created, there is no harm in *singling out one* of them by means of a property involving *all* of them. Poincaré consequently held this kind of circularity to be non-vicious. For him but not for Russell, 'the tallest man in this room' is a legitimate description; for it is meant to characterize rather than to construct the designated entity. In this way, we obtain a variant of Poincaré's (VCP)—denoted by $P1_a$—which Heinzmann formulates as follows:

> A definition [i.e. determination] of a set e involving in its definiens a totality E of which e forms part is nonetheless predicative if 'e is not part of E by virtue of its definition, but as a result of a proof which precedes the definitions of both E and e'. (Heinzmann 1985, p. 62; my translation)

Let us incidentally note that by contrast with both Russell and Poincaré, a thoroughgoing realist regards the entities referred to in a mathematical proof as being given prior to the actual deduction; whereas according to [P], a mathematical object exists only if it can be constructed in the pure intuition of time. Such constructions consist of series of operations some of which are classifications of previously defined entities. According to Poincaré, every class M is the result of *effectively* dividing a number of constructed objects into those which fall within, and those which lie outside M. An allowable entity will be defined by this process only if M eventually becomes stable; i.e. if posterior classifications do not overturn antecedent ones. More generally, beyond a certain point in time, an intended class should remain unaffected by all subsequent definitions. Should a classification be continually altered, then all references to *the* set generated by it become null and void; moreover, any object dependent on such an illegitimate collection would keep being modified during the constructive process and thus become indeterminate.

> The rules of formal logic simply express the propertirs of all possible classifications. But in order that they should be applicable, it is necessary that these classifications should be immutable and not require to be modified in the course of the argument. If we have only to classify a finite number of objects, it is easy to preserve these classifications without change. If the number of objects is indefinite, that is to say if we are constantly liable to find new and unforeseen objects springing up, it may happen that the appearance of a new object will oblige us to modify the classification, and it is thus that we are exposed to the antinomies. (Poincaré 1908, pp. 194–95)

This stability or immutability condition, which is to be imposed on all classifications, gives rise to a version of the (VCP) which Heinzmann denotes by P_2:

> Every correspondence rule presupposes a classification. I call a correspondence rule predicative if the presupposed classification is predicative. But I call a classification predicative if it is not overturned by new elements.
>
> This is not the case with Richard's classification: the introduction of the correspondence rule overturns the classification of all sentences in terms of whether they possess, or fail to possess, a meaning. (Heinzmann 1985, p. 70)

Poincaré's immutability requirement can be illustrated as follows. Let OT denote the time axis; and let t_1, t_2, $\underline{t}$, t_1^* and t_2^* be the abscissae of five points of OT such that $t_1 < t_2 \leq \underline{t} < t_1^* < t_2^*$. Suppose that a classification, carried out at time t_1, gives rise to some class E; and that at t_2, an object *a* is constructed by means of an operation Q which might involve E. Thus both E and *a* can be said to exist at t_2. Assume further that at t_1^*, E is altered into E* by the adjunction to E of some newly created elements such as *a*; where, as in the case of *a* itself, the construction of these new entities might have involved E. As a consequence of the alteration of E into E*, *a* will generally be changed into *a** at some instant $t_2^*>t_1^*$; where *a** is the result of carrying out the operation Q* obtained by substituting E* for E in the wording of Q. Generally speaking, both E* and *a** will exist at t_2^*, at which point one of three very different situations might arise:

[1] *a** exists and *a** = *a*. This is the unproblematic case.

[2] *a** exists, but *a** ≠ *a*.

[3] *a** does not exist either because E* cannot be constructed or because Q* is unperformable.

If, beyond every stage $\underline{t} \geq t_2$, there exist $t_2^*>t_1^* \geq \underline{t}$, such that at t_2^*, either Situation [2] or Situation [3] arises, then the definite description 'the object *a* obtained by method Q' is judged by Poincaré to be illegitimate: it fails to denote anything because it possesses either too many putative referents or none at all. Antinomies are then bound to arise, for by using phrases like 'the *a* obtained through applying Q', we confuse many differently defined entities some of which may not even exist. For instance, in Situation [2], we might be misled into writing *a* = (object defined by means of Q) = (entity defined by means of Q*) = *a**, from which, given *a*≠*a**, a contradiction follows.

Poincaré's immutability condition states that there should be a stage $\underline{t} \geq t_2$ such that: both the corresponding class $\underline{E}$ and the object $\underline{a}$ exist at $\underline{t}$, and $a^* = \underline{a}$ for all $t_2^* > t_1^* \geq \underline{t}$. We could then legitimately stipulate that the expression 'the object defined by means of Q' should exclusively denote the entity $\underline{a}$.

Let us examine more precisely the way in which Principle [P] excludes circular definitions. Consider an object *a* which is to be defined in terms of a class E; where some of E's members are, in their turn, supposed to have been defined by means of *a*. In the light of [P], it is easily seen that neither *a* nor E could possibly exist; i.e. neither entity is constructible in the pure intuition of time. For since the construction of *a* involves E, the definition of *a* cannot precede that of E. Similarly, E cannot precede *a* because some elements of E require the prior existence of *a* in order to exist. Thus the notions of E and of *a* violate the linearity of the ime-ordering relation; so that in accordance with [P], neither *a* nor E is a legitimate mathematical object.

Put in a slightly different form, the above argument can also be expressed as follows. We have just seen that *a* cannot precede E, for *a* needs the prior definition of E in order to exist. E must therefore be taken to have been determined before we can deal with *a*. But the only set available before the construction of *a* is not E, but the truncated class E_0 obtained by eliminating from E all elements presupposing *a*. Thus $E_0 \subset E$ (strictly). Suppose that some object *a* can nonetheless be obtained by operating on E_0 instead of E. Once constituted, *a* automatically generates the set $E \setminus E_0$ whose elements arise from the existence of *a*. E_0 is thus instantaneously completed: it is transformed into E by the presence of *a*. The creation of the entity *a* overturns the class E_0 into E, so that *a* must now be redefined by operating on E instead of E_0. There is however no guarantee that this new definition will yield the same *a* as the one previously constructed (see case [2] above). Whence the following paradox: the presence of *a* might undo *a* itself; which is but another way of saying, by way of a *reductio*, that this whole process of construction is impossible.

Let us apply this kind of reasoning to Berry's paradox where a logical contradiction would arise if *a* could be properly defined as the *smallest natural number greater than every number nameable in less than one hundred words.* Let S be the list of all sequences of less than 100 English words. S is not only a finite set but through the use of a comprehensive dictionary, it can be effectively enumerated. Let H be the subsequence of S consisting of all well-formed *terms* each of which singles out a natural number. True to his constructivist stance, Poincaré required that we be able to exhibit the number denoted by any given expression belonging to H. H is thus obtained by eliminating from S all phrases which do not effectively name a definite number. One could start by striking out all sequences of symbols in S which are not syntactically well-formed. One would then eliminate all syntactically correct expressions which constitute sentences—as opposed to terms. For the sake of argument, suppose that being based on purely syntactical criteria, these two operations can be actually carried out. One would thus be left with a finite sequence of well-formed terms. We must now eliminate all those terms which either do not denote numbers or fail *effectively* to refer to anything at all. This operation, which possesses a semantic aspect, may prove unperformable; whence the expression 'semantic paradoxes'.

Should H turn out to be constructible, then E could be defined as the set of all natural numbers named by elements of H. Thus: $E = ||H||$ where $||H||$ denotes the class of all objects signified by members of H. (Remember that H is a collection of syntactical expressions.) *a* can finally be computed as $\text{Max}(E) + 1 = \text{Max}(||H||) + 1$. *a* is thus definable iff $||H||$ can be effectively circumscribed; or equivalently: iff H can be constructed. Let us show that *a* cannot be properly defined. Suppose otherwise; i.e. assume that all the elements of H can be written down. Let **Q** stand for the expression 'the smallest natural number greater than every natural number nameable in less than one hundred words'. Since **Q** itself contains less than 100 words, then by definition of S, **Q** $\in$ S. In order to enumerate all the members of H, we must therefore have decided whether or not **Q** is to be struck out from S; i.e. whether or not **Q** effectively denotes a number; i.e whether or not **Q** belongs to H. In other words, we must already have written down all the elements of H. Of the two operations: *defining* H and *deciding whether* **Q** *should be eliminated from* S, each ought therefore to precede the other. This contradicts the asymmetry of the temporal relation and hence entails that neither H nor a = Max(||H||) + 1 can be determined; i.e., neither entity exists.

The above argument can be reformulated in a way which perspicuously brings out the connection between the antinomies on the one hand, and the instability of certain classifications on the other. Before the members of H have all been written down, the expression **Q**, which is supposed to denote Max(||H||) + 1, possesses no referent. By definition of H, it follows that **Q** $\notin$ H. Assume that H can eventually be completed in time. As soon as this is done, **Q** automatically acquires the number q_1 = Max(||H||) + 1 as effective referent and will therefore lie in H. Since **Q** $\in$ H, then by definition of ||H||, $q_1 \in$ ||H||. So the referent of **Q** has by now been turned into q_2 = Max(||H||) + 1 > q_1. Since q_2 exceeds every member of the old ||H||, q_2 *could not have belonged* to ||H||; but it *now does*, since q_2 is the *new* referent of **Q** $\in$ H. But this entails that **Q** must denote yet another number, namely q_3 = Max(||H||) + 1 > q_2; and so on indefinitely. In other words: neither the referent of **Q** nor the extension of the class ||H|| can be stable over time: for **Q** successively denotes each member of the increasing sequence q_1, q_2,...; while ||H|| keeps acquiring the q_i's as new elements. Since the immutability condition fails, then according to Poincaré, the referent of **Q** does not exist and the paradox consequently vanishes.

Vicious circularities may therefore upset some classifications and prevent the construction of entities which seem *prima facie* unexceptionable; and the Berry paradox shows that the sets destabilized in this way need not be infinite. This is why Russell rightly drew Poincaré's attention to the fact that the antinomies need not involve any actual infinities. Yet Poincaré went on maintaining that vicious circularities have to do with extending to infinite sets operations which make sense only when applied to finite collections, and he would be vindicated if it could be shown that infinities give rise to class instability in situations where finite collections would cause no serious problems. I shall therefore allow myself to construct the following imaginary example.

Let B = $\{b_1, b_2, ...\}$ be a denumerably infinite set which a Cantorian would regard as a completed totality; and let A be any class about whose cardinality we need make no assumptions. The Cantorians consider themselves entitled to envisage a relation R between any element x of A and the whole of B. This leads to a partition of A into the two subclasses A_1 and A_2 defined as follows:

(27) $A_1 = \{x \in A: xRB\}, A_2 = \{x \in A: \neg xRB\}$.

For a pragmatist like Poincaré, B does not however exist as a completed entity but as the virtual limit of a process which successively defines the following finite sets:

(28) $B_0 = \emptyset, B_1 = \{b_1\}, B_2 = \{b_1, b_2\},..., B_n = \{b_1,..., b_n\},...$; so that

(29) $B = \cup\{B_n: n = 0 \rightarrow \infty\}$, where ∞ is never actually reached.

For the sake of argument, let us assume that the construction of the finite collections B_0, B_1,...poses no serious problems; more particularly, that their definitions involve no vicious circularity. (Even though each B_n be finite, Berry's paradox shows such an asumption to be far from innocuous). We shall see that the infinite character of B may, even then, destabilize certain classifications.

The pragmatist regards the predicate R, which purports to relate an object x to the actually infinite class B, as illegitimate. He will consequently try to define a series of relations R_0, R_1,... such that: for each $n \in \omega$, each $x \in A$ and every set Y consisting of n members, xR_nY is a pragmatically meaningful statement. For large n, 'R_n' is supposed to 'tend asymptotically' to the *intuitive* meaning of 'R'. Each R_n will thus give rise to the following partition of A:

(30) $A_{n1} = \{x \in A: xR_nB_n\}, A_{n2} = \{x \in A: \neg xR_nB_n\}$.

Thus, for each n:

(31) $A = A_{n1} \cup A_{n2}, A_{n1} \cap A_{n2} = \emptyset$; but we may also generally have:

(32) $A_{n1} \neq A_{m1}$, hence also $A_{n2} \neq A_{m2}$, for $n \neq m$.

In order to reconstruct what the Cantorians mean by the partition $<A_1, A_2>$, the pragmatist will want to say that ultimately, i.e. for sufficiently large n, the partition $<A_{n1}, A_{n2}>$ becomes immutable; in other words: for some $p \in \omega$ and for all n and m such that $n \geq p$ and $m \geq p$, $<A_{n1}, A_{n2}> = <A_{m1}, A_{m2}>$ holds. This would enable the pragmatist to identify $<A_1, A_2>$

with *any* $<A_{n1},A_{n2}>$, where $n \geq p$. He must therefore presuppose the existence of some $p \in \omega$ such that:

(33) $(\forall n \geq p)[(A_{n1} = A_{p1}) \wedge (A_{n2} = A_{p2})]$.

By (30), it follows that:

(34) $(\forall n \geq p)(\forall x \in A)[(xR_nB_n) \Leftrightarrow (xR_pB_p)]$; we must consequently assume that

(35) $(\exists p \in \omega)(\forall n \geq p)(\forall x \in A)[(xR_nB_n) \Leftrightarrow (xR_pB_p)]$.

(35) need not however be satisfied for every (appropriate) R. Thus (33) might fail for every p, i.e. $(\forall p \in \omega)(\exists n \geq p)[A_{n1} \neq A_{p1}]$ might be true. This means that beyond every stage p, there exists a stage n which *overturns* the classification $<A_{p1}, A_{p2}>$ in the sense that: $<A_{n1}, A_{n2}> \neq <A_{p1}, A_{p2}>$. In such a case, a pragmatist would declare the partition $<A_1, A_2>$ illegitimate; he would also add that these difficulties arise because B is infinite; for otherwise, R would, right from the start, be a well-defined relation between an object x and the *finite* set B. Poincaré was therefore right in thinking that the problem of class instability is *made worse* by the alleged existence of actual infinities.

5.6 Poincaré, Russell, and Modern Empiricism

[A] The Vienna Circle

In Chapter 3 we discussed the profound influence which Poincaré's views about the status of geometry had on those of eminent empiricists like Carnap, Reichenbach, and Grünbaum. When we turn to the foundations of set theory, that is, of mathematics proper, the picture becomes somewhat blurred. Russell is generally regarded as one of the founding fathers of the neopositivist movement. Yet Schlick initially shared Poincaré's quasi-intuitionist position against Frege's and Russell's logicism. He wrote:

> This view concerning mathematical certainty to which Leibniz and Hume had also come close, and according to which mathematics can in some sense be reduced to logic, was vehemently opposed by Poincaré. His argument consists in denying that in mathematics, one can dispense with all forms of induction. . . . As regards this inner process, we completely share Poincaré's opinion. . . . He rightly considers any proof carried out by means of an inference from n to n + 1 as the "raisonnement mathématique par excellence" (the kernel of mathematical inference); and he

> refuses to call it 'complete induction', a name which is usually given to it; for one does not even think of actually carrying out the proof for all the numbers. . . . How can one explain, according to Poincaré, this apodictic certainty which is inappropriate to other forms of induction? "For it is only the confirmation of a mental capacity convinced of being able to represent to itself the repetition of one and the same step, once a single such step is known to be possible." (Schlick 1986, pp. 84–85; my translation)

Schlick thus started out by being a Kantian who supported Poincaré's views not only about the status of geometry, but also about the foundations of mathematics in general. Under Wittgenstein's influence, he then underwent an anti-realist conversion and turned into a hard-nosed empiricist. Because of Wittgenstein's perplexing talk about mathematics not consisting of any genuine propositions but of grammatical rules, Schlick ended up maintaining that:

> The fact is that numbers or other logical entities cannot be regarded as 'objects' in any sense at all. The propositions apparently dealing with them do not communicate any 'facts' about them and therefore are no 'propositions' at all. What are they?
>
> They are nothing but certain rules which determine the use of language, i.e. of expression by combination of 'signs'. They are concerned with symbols, not with reality. (Schlick 1938, pp. 145–46)

This passage lays bare the essential weakness of the neopositivists' appraisal of both logic and mathematics. The Logical Empiricists put forward two theses: first, that a proposition p should be considered meaningful iff it is in principle verifiable; secondly, that p and ¬p are either both meaningful or both meaningless. The second thesis is clearly ungainsayable; but the first one—which conflates the meaning of a statement with the method of its verification—proves totally wrong-headed. First, it threatens to ban as meaningless all universal and hence all scientific hypotheses. Secondly and just as seriously, it entails that logical truths are 'senseless'; for if t is a tautology, then ¬t cannot conceivably be verified; so ¬t, and hence also t, must be considered devoid of meaning. It is however difficult to accept that (p∨¬q), which expresses a normal proposition as long as p is distinct from q, should suddenly becomes senseless when p≡q. Ironically, the formula (p∨¬p), which was declared empty and hence meaningless by both Wittgenstein and Schlick, proved far from being an *innocuous* tautology: as is well-known, the intuitionists regard (p∨¬p) as a meaningful proposition but they deny its universal validity.

The neopositivists actually oscillated between the extremist thesis described above and the more sensible view according to which tautologies are limiting cases of genuine propositions; i.e. formulas whose truth-value turns to be the 'True' no matter how their descriptive terms are interpreted. Wittgenstein and Schlick noneteless continued to consider logical truths involving the identity symbol, for instance $[(x = y) \rightarrow (H(x) \leftrightarrow H(y))]$, not as genuine statements but as rules allowing us, whenever we accept $(x = y)$, to replace x by y in any formula H(x).

So far, we have dealt mainly with logical problems. When we turn to the foundations of mathematics, the situation becomes more complex and even less satisfactory. Some members of the Vienna Circle like Carnap and Hahn took all mathematical truths to be tautologous. According to Rudolf Carnap,

> Mathematics, as a branch of logic, is also tautological. In the Kantian terminology: the sentences of mathematics are analytic. They are not synthetic *a priori.* (Ayer 1959, p. 143)

It should be noted that Carnap wrote these lines in 1930–1931, nearly twenty years after Poincaré's death. It was as though the Poincaré-Russell debate which culminated in the recognition—by both protagonists—of the synthetic character of mathematics, had never taken place. Hans Hahn more guardedly maintained that:

> To be sure, the proof of the tautological character of mathematics is not yet complete in all details. This is a difficult and arduous task; yet we have no doubt that the belief in the tautological character of mathematics is essentially correct. (Ayer 1959, p. 158)

Thus Hahn seems to have been vaguely aware of the difficulties confronting the logicists, but he did not doubt the latter's eventual success. He ought however to have realised that Russell himself acknowledged the synthetic character of mathematics; for as seen in the previous sections, Russell not only viewed his own Axiom of Reducibility with great suspicion but also conceded that the axioms of Infinity and of Choice are contingent.

Turning once more to Wittgenstein and Schlick, we find them blurring this issue by claiming that like all logical principles, mathematical formulas do not express genuine propositions but are grammatical rules governing the interchangeability of syntactical expressions. According to the later Schlick, there is in fact an absolute excluded middle between experience and grammar:

> '5 + 7 = 12' is no proposition at all, it is a rule which permits us to transform a proposition, in which the signs 5 + 7 occur, into an equivalent one in which the sign 12 occurs. It is a rule about the use of signs and therefore does not depend on any experience, but only on the arbitrary definitions of the signs. (Schlick 1938, p.222).

> Similarly to arithmetic, geometry does not refer to any objects at all, e.g. to points, lines or surfaces; it consists of rules telling us what measurements and constructions are to be deemed permissible. For example: the acceptance of Euclidean geometry enables us to claim that if $\alpha + \beta + \gamma$ sensibly differs from π, then our measurements must be at fault; where α, β and γ are the measured angles of any triangle. (McGuinness 1967, pp. 61–63, 161–63).

Thus, no matter how divergent their views about the foundations of mathematics might have been, the neopositivists agreed on one fundamental tenet: mathematics is contentless in that its postulates—even when conceived as rules—single out neither a unique model nor even a class of models; they could be about anything and everything; they formed a mere organon for the derivation of empirical propositions from contingent assumptions. That the Vienna Circle's conception of the status of mathematics is totally misguided can be established by a Popperian type of transcendental critique, that is: by a criticism based on the examination of the methods successfully used within the sciences (Popper 1979, Chapter 4, §9). To the extent that these methods prove incompatible with the neopositivistic theses, the latter will have been undermined; or, more modestly: the onus of proof will have been shifted onto the logical empiricists whose views are then seen to conflict with well-tested scientific practice. As a matter of fact:

(1) Even a cursory look at any scientific system shows that both the mathematical and the logical principles are treated, not as rules, but as ordinary propositions to which *deductive rules are applied.*

(2) *All* the mathematical axioms cannot possibly be tautological: apart from the anti-logicist conclusions reached by Poincaré and Russell, Boolean-valued models were more recently used in order to falsify not only such esoteric postulates as the Generalised Continuum Hypothesis (GCH) but also innocuous-sounding principles like the Axiom of Choice (AC). These mathematical postulates were thus shown to be synthetic; provided, of course, that the remaining axioms of set theory be consistent; but if they are not, then mathematics would be self-contradictory and hence certainly non-analytic (see Bell 1977).

(3) Not only are the postulates of mathematics logically independent of the remaining components of any scientific system, but the latter can nor-

mally be formulated only against some pregiven mathematical background. Hence the relationship between mathematics and logic on the one hand, and physics on the other, is asymmetric.

(4) Thus the global language of most scientific hypotheses contains both mathematical and physical symbols. By virtue of this linguistic entanglement, some mathematical axioms can be brought to bear on physical propositions. This leads to the question whether the ingression of mathematics into physics might not cause an increase in the empirical content of scientific hypotheses.

(5) Kant's thesis that mathematical theorems express synthetic truths thus seems vindicated. But truths about what? It proves impossible to give a physical interpretation to *all* the mathematical terms occurring in a scientific system. Do such terms have any referents at all, and if so, are we not thereby driven to accept a Platonist form of realism as the only coherent position?.

In what follows, points (2)–(5) will be discussed in some detail.

[B] The Problem of Mathematical Realism within The Empirical Sciences

Quine claims that whenever we quantify over a variable, we are implicitly committed to the existence of the objects over which the variable ranges (Quine 1980b). As explained above, mathematics, and more particularly set theory, are systematically applied to the physical sciences. This is why we often have to quantify over classes whose existence must, according to Quine, be presupposed. When used within a scientific system, set theory thus entails the existence of sets, that is: of abstract entities.

I find Quine's thesis, *to be is to be quantified over*, seductive but not altogether compelling; for nothing prevents us from regarding classes as convenient fictions, or from ascribing to statements like 'the empty set exists' a figurative rather than a literal sense. Such a move would however leave one major problem unsolved. Physical theories are supposed to be about the real world; and at least some of their statements, namely the observation reports, definitely possess referents which make them true-or-false. Note furthermore that logical principles are applied throughout a scientific system: to the physics, to the underlying mathematics and to any mixture of the two. The main rationale behind the use of 'logic' is the transmission of truth and hence the retransmission of falsity. Thus an important reason for obeying the logical rules, even when dealing with allegedly abstract mathematical statements, is that the latter must have some truth-value to transmit and consequently possess referents. We have also pointed out that many set-theoretic axioms are synthetic. In other words, they could turn out false under certain interpretations; they must therefore be true-or-false of *something*. Again, the problem lies with the nature of this *something*.

According to Platonism, mathematical terms refer to Forms dwelling in an aetherial realm. We might, in the end, have to concede that this is the most coherent conception of the nature of mathematics; but most modern realists would opt for some form of Aristotelianism rather than for full-blown Platonism. Let me define what, in the present chapter, will be meant by Aristotelianism or rather by quasi-Aristotelianism. Our view of the world is obviously informed by theoretical expectations. We moreover possess a well-developed capacity for abstraction and idealisation, that is: an ability to simplify or screen off some aspects of the messages conveyed by the environment. Such an ability has no doubt great survival value: if we tried to attend to all the details of the world's input, then we might be unable to see the wood for trees; we would be paralyzed by the sheer wealth of the information received and consequently perish as a species. This is why we see—or imagine that we see—global patterns which we then proceed to idealize; in other words we purify these structures and isolate them from their concrete setting in order more effectively to study their properties in a discipline misleadingly called pure mathematics. We finally reintegrate the systems thus constructed into the physical context from which they were extracted; it is then found that the mathematical results obtained in this way apply to situations for which they were not initially intended. According to Aristotelianism—as here understood—all mathematical forms are embodied in the world from which, for reasons of mere convenience, they are artificially detached. Let me give a few examples which support this doctrine. Right from the start, Greek geometry was meant to be about real—albeit idealized—space; it was a study of pure extension which abstacted from such empirical attributes as colour, smell and sound; and when, in the nineteenth century, Riemann put forward his new conception of geometry, he based it on an idealized notion of the *measurement* of small distances. It is therefore hardly surprizing that geometry should be applicable to physics. Another example is provided by the Differential Calculus which owes its origins both to mechanics and to analytic geometry. Newton looked upon the derivative as a fluxion, i.e. as a velocity, and upon time as the universal fluent. As for Leibniz, he viewed the differential coefficient as representing the direction of a smooth graph. These results were achieved independently of any *foundational* analysis of the new mathematics, which was not carried out until the nineteenth century. Thus the Calculus would have been neither discovered nor further developed, had its first proponents failed to look upon derivatives and integrals as properties of concrete paths defined over the continuum. It was therefore natural to regard both $\aleph_{0,}$ i.e. the cardinal of the set ω of natural numbers, and $\exp(2,\aleph_0)$, i.e. the cardinal of the real line **R**, as being actualised in the physical world; for Newton's temporal fluent and Einstein's 4-dimensional world, both of which have $\exp(2,\aleph_0)$ as their cardinality, can be taken to form part of physical reality (Henceforth, for any two cardinals h and k, the cardinal h^k will often be denoted by exp(h,k).)

Let us now turn to set theory. As already mentioned, 'abstraction' describes the process of screening off, or of omitting certain items of information. All observed classes are obviously finite. Yet some notions like those of equinumerosity, of the power set or of the union of a family of sets can be defined independently of the cardinality of the underlying classes; i.e. the size of these classes need not be mentioned at all. This is precisely what led Galileo to wonder whether the sequence of positive integers has as many members as one of its subcollections, namely that of all even numbers. A considerable part of set theory can therefore be formulated agnostically, without any presuppositions as to the size of its objects. Only when one comes to reconstruct analysis does the need to posit explicitly the existence of an infinite set like ω arises. Let us recall that $|\omega| = \aleph_0$ and that both $\aleph_0$ and $\exp(2,\aleph_0)$ can be taken to be physically exemplified; hence these two infinite cardinals seem to raise no difficult conceptual problems. All high-level hypotheses however have unexpected consequences; so that from set theory we obtain much more than we initially bargained for. Let us put $c_1 = \exp(2,\aleph_0)$ and for all natural numbers j, inductively define: $c_{j+1} = \exp(2,c_j)$; by Cantor's theorem: $c_j < c_{j+1}$; hence: $\aleph_0 < c_1 < c_2 < c_3 <$... Although we might hold $\aleph_0$ and c_1 to be spatio-temporally embodied, we find it hard to believe that c_2 or c_3 can be physically realized. So why not simply ignore these monstrously large entities? The problem is that the unexpected consequences of set theory are not all unwelcome. In physics, we need to prove the existence of solutions to certain differential equations while being unable to express these solutions in terms of any known functions. That is, we often have to content ourselves with purely existential theorems of the form: $(\exists f)[\mathbf{F}(f) \wedge \mathbf{E}(f) \wedge \mathbf{B}(f)]$, where: $\mathbf{F}(f) \equiv$ f is a function, $\mathbf{E}(f) \equiv$ f satisfies the equation **E**, and $\mathbf{B}(f) \equiv$ f satisfies the boundary conditions **B**. More importantly, even in the down-to-earth empirical sciences, we often have to envisage so-called 'pathological' mathematical entities. For example: while studying heat phenomena, Fourier introduced into analysis a host of discontinuous functions qua limits of certain trigonometrical series. Modern Quantum Mechanics explicitly deals with discontinuous processes; so it would be highly undesirable to exclude from its mathematical methods all 'pathological' constructs. In probability theory, one often has to resort to the Lebesgue integral which allows into mathematics a whole class of 'abnormal' functions. Some variables like f above will therefore be required to range over a *set G* encompassing not only all well-behaved functions but also arbitrary real-valued maps. The cardinality of *G* is: $\exp(c_1,c_1) = \exp(\exp(2,\aleph_0),c_1) = \exp(2,\aleph_0\, c_1) = \exp(2,c_1) = c_2$. Note finally that Quantum Physics involves the use of functionals and of operators which send functions either into numbers or into other functions. We are thus led to consider domains of cardinality: $\exp(c_1,c_2) = \exp(\exp(2,\aleph_0),c_2) = \exp(2,\aleph_0\, c_2) = \exp(2,c_2) = c_3$, or $\exp(c_2,c_2) = \exp(\exp(2,c_1),c_2) = \exp(2,c_1c_2) = \exp(2,c_2) = c_3$.

We therefore face the problem of positing the existence of entities like c_2 and c_3 which, despite being defined by meaningful mathematical propositions, cannot reasonably be held to have any physical representation. The fact that $\aleph_0$ and c_1, but not c_2 or c_3, are physically exemplified supports A.N. Whitehead's claim that there is an ingression of some, but not of all the Platonic Forms into the material world (Whitehead 1926, Chapter 10).

[C] Hartry Field's Dismissal of Mathematical Ontology

As already mentioned, after Cohen's proof of the independence of (AC) and of (GCH) from the remaining postulates of set theory, the challenge posed by the synthetic character of mathematics could no longer be ignored. One had either to accept Poincaré's Kantian thesis that mathematics is in part synthetic *a priori*; or Quine's view that mathematics is in principle *revisable*; or finally that despite being synthetic, mathematics is in some sense dispensable. Of these alternatives, Hartry Field opted for the last (Field 1980).

According to Field, mathematics is useful in that it enables us to simplify certain proofs or at least to render them more perspicuous; but when it comes to deriving the *nominalistic* or *non-abstract* results which Field regards as the only ones possessing any genuine physical meaning, all mathematical axioms allegedly turn out to be dispensable; they provide convenient but strictly superfluous premises; for they prove to be conservative over all nominalised versions of scientific theories (see Field 1980, Appendix to Chapter 1). Let me adduce a parallel which clarifies Field's thesis. It is well known that despite being invoked in order to simplify some mathematical deductions, the Axiom of Choice (AC) can be circumvented in many proofs, for instance in that of the classical Heine-Borel theorem. According to Field, *all* mathematical assumptions are similarly used in order to simplify the derivations of *nominalistic* scientific results. The latter however in no way presuppose these assumptions. Mathematical ontology might consequently be of interest to the pure mathematician, but it need not worry the physicist; for science can in principle do without any mathematics at all.

Field's undertaking can be viewed as a last-ditch attempt made by modern empiricism to show that applied mathematics is, if not logically, then at least empirically empty. In this way, we retrieve the Vienna Circle's contention that all substantive knowledge is a posteriori, while logic and mathematics constitute mere modes of presentation of that knowledge.

[D] Field's Results and His Method of Proof

In Chapters 1–4, I tried to vindicate Poincaré's thesis concerning the *synthetic a priori* character of mathematics against the positivist excluded middle between the empirical and the analytic. It is therefore essential for us to examine Field's results and the soundness of his proofs.

Following Field's and Chihara's own accounts, let me describe—in formal terms—the way in which mathematics is generally thought to impinge on scientific knowledge (Chihara 1990, Chapter 8).

Let **T** be any hypothesis whose vocabulary will henceforth be denoted by **V**; and let $\mathbf{V'} = \mathrm{V} \cup \{\in, s, N\}$; where $\in$ and s are as described on p. 000, while N is a monadic predicate whose intuitive meaning is given by: $N(x) \equiv$ x is a nominalistic, in other words a non-abstract or non-mathematical, entity.

Now define **T'** as the hypothesis which is expressed in the enlarged vocabulary **V'** and obtained by adjoining to **T** the mathematical theory $\mathbf{S_{V'}}$; where the axioms of $\mathbf{S_{V'}}$—also expressed in the language **V'**—consist of the following three groups:

(α) $(\forall x)[s(x) \rightarrow \neg N(x)]; \exists(x)[s(x) \wedge (\forall y)(y \in x) \leftrightarrow N(y))]$.

(β) This group consists of the single axioms **[1]–[4]** and **[7]–[11]** of (ZFC) (see pp. 000–000). All these axioms can be expressed exclusively in terms of the logical constants and of the primitives: s and $\in$.

(γ) This group consists of the Axiom-*Schemes* of Replacement and Separation, i.e of **[5]** and **[6]** (see p. 000). Let us point out that in (α) and (β) above, every set is meant to comprise either mathematical entities *or* physical objects *or* both. Yet our theories **T** and $\mathbf{S}_{\mathbf{V}'}$ seem to be disjoint; for apart from the logical constants, $\mathbf{S}_{\mathbf{V}'}$ appears so far to have only s, ∈ and N, as primitive symbols. The *Schemes* **[5]** and **[6]** remedy this defect by building bridges between the mathematical and the physical components of the scientific system **T**'; for *E(*x*)* and *C(*x,y*)* now represent arbitrary wffs of the extended language **V**'. (It should again be noted that the Separation Scheme is redundant: it follows, without the use of the Axiom of Choice, of the Axiom of Foundation, or of the Continuum Hypothesis, from the remaining axioms and axiom schemes of (ZF).)

Without any loss of generality, it can be assumed that **V**' contains no function symbols and hence no names. For reasons of convenience, we may however allow the introduction of finitely many individual names $a_1, a_2, \ldots, a_p$; the only proviso being that for each j in {1,2,...,p}, either $N(a_j)$ or $\neg N(a_j)$ be specified as an axiom of **T'**. We might for example find it convenient to denote the empty set by the constant ø; and in Quantum Mechanics, the underlying Hilbert space might be referred to as *H*. In this case, ¬N(ø) and ¬N(*H*) would have to be added as axioms; for both ø and *H* are classes and hence (abstract) mathematical objects.

We shall subsequently need the following *definition*: a name a_j will be called either nominalistic or else abstract according to whether $N(a_j)$ or $\neg N(a_j)$ is an axiom of **T**'. And a wff *B* of **V**' is said to be nominalistic iff: *B* contains neither any abstract names nor the predicates s or ∈; and if, furthermore, all the bound variables occurring in *B* range over nominalistic entities; i.e. every quantifier of *B* occurs in one of the following contexts: $(\forall x)[N(x) \rightarrow \ldots]$ or $(\exists x)[N(x) \wedge \ldots]$.

For any wff *H*—which is normally taken to contain neither s nor ∈ nor any abstract names—the nominalized counterpart *H** of *H* is obtained from *H* by successively replacing—say from right to left—all subformulas of the form $(\forall x)B(x)$ and $(\exists x)B(x)$ by $(\forall x)[N(x \rightarrow (B(x)]$ and $(\exists x)[N(x) \wedge B(x)]$ respectively. It is finally worth noting that the property of being a nominalistic formula is (syntactically) decidable.

Field establishes the following conservativeness theorem. Let Λ be any collection of nominalistic postulates such that $\mathbf{S}_{V'} \cup \Lambda \vdash g$, where g is a nominalistic sentence; then $\Lambda \vdash g$. Thus the mathematical theory $\mathbf{S}_{V'}$ though possibly useful, is not indispensable to the derivation of g from Λ.

In order to find what extra assumptions Field needs to establish this result, we ought to examine his proof, which runs—roughly—as follows. Suppose that $\mathbf{S}_{V'} \cup \Lambda \vdash g$ but Not($\Lambda \vdash g$). $\Lambda \cup \{\neg g\}$ is therefore consistent. If it could be shown that $\mathbf{S}_{V'} \cup \Lambda \cup \{\neg g\}$ is also consistent, the assumption $\mathbf{S}_{V'} \cup \Lambda \vdash g$ would have been contradicted and the theorem proved (by *reductio*). Let us restrict ourselves to first-order languages. Since $\Lambda \cup \{\neg g\}$ is consistent, it must have some model, Σ say. By making sure that Σ satisfies certain conditions, by then applying iterative methods up to the first inaccessible cardinal, Field shows how to extend Σ to the whole of $\mathbf{S}_{V'} \cup \Lambda \cup \{\neg g\}$. The consistency of this last set of sentences is thereby established. Note that the consistency of $\mathbf{S}_{V'}$ and hence that of set theory would thereby also have been proved. Given Gödel's second incompleteness theorem, it is hardly surprising that Field needs to add, at the meta-level, postulates going beyond $\mathbf{S}_{V'}$; for example, an axiom about the existence of inaccessible cardinals or the assumption that set theory is ω-consistent. Such presuppositions could justifiably be considered as too strong. It should nonetheless be granted that as regards first-order nominalistic theories, Field has succeeded in replacing the question of mathematical truth by that of—strong—mathematical consistency. It is however also fair to point out that the cost is high.

[E] The Ultimate Failure of Field's Programme

Though the conservativeness theorem is an interesting and ingenious *mathematical* result, it does not solve the fundamental problems raised by nominalism; for a scientific theory is normally presented, not as a series of nominalistic postulates, but as an already mathematicized hypothesis such as **T'**, which possesses a rich surplus structure. Given any such **T'**, Field faces the task of constructing a nominalistic alternative **G** having the following properties:

(1) For any nominalistic sentence g, $\mathbf{T'} \vdash g$ entails $\mathbf{M} \cup \mathbf{G} \vdash g$; where **M** is an appropriate set of mathematical axioms. E.g., **M** could be the set-theoretic system $\mathbf{S}_{V'}$ described above.

(2) **M** is nominalistically conservative over **G**. More precisely: for any nominalistic g such that $\mathbf{M} \cup \mathbf{G} \vdash g$, we also have $\mathbf{G} \vdash g$.

(3) As an instrument, **M** must finally prove to be, not only expendable, but also useful.

In order to establish (3), Field proved a so-called representation theorem, a proposition which enables him to go from a nominalistic g to its abstract counterpart g'; then to bring powerful mathematical techniques to bear on g', thus drawing some abstract conclusion h'; finally to cross from h' back to a corresponding nominalistic proposition h. Invoking the conservativeness clause (2), one would then be in a position to assert $\mathbf{G} \vdash (g \rightarrow h)$.

Should **G** be merely required to be an axiomatizable nominalistic theory satisfying (1)–(3), then we should have to concede that Field's undertaking was bound to succeed. All Field needs to do is to adopt the well-known Craigean approach to such problems. Let me give a brief description of Craig's methods, if only in order to show why Field understandably rejected them. **T** is usually presented as a recursively axiomatized hypothesis, which is also the case of the mathematical theory $\mathbf{S}_{V'}$. It follows that **T**', i.e. the consequence class of $\mathbf{T} \cup \mathbf{S}_{V'}$, can be axiomatized; it therefore consists of a recursively enumerable sequence $\{B_0, B_1, ...\}$ of formulas. Let us recall that the property of being a nominalistic sentence is decidable and that there exist infinitely many syntactically distinct nominalistic elements of $\{B_0, B_1, ...\}$; for if we put $K = (\forall x)[N(x) \rightarrow N(x)]$, then K, being logically true, must lie in $\{B_0, B_1, ...\}$. It follows that $K^{(n)} \in \{B_0, B_1, ...\}$ where, for any wff Y, $Y^{(n)}$ stands for the conjunction $(...(Y \wedge Y) \wedge ...Y)$, in which n is the number of distinct occurrences of the conjunct Y. (For any Y and any positive integer n: $\vdash (Y \leftrightarrow Y^{(n)})$.)

Let Σ be the set of all formulas X of the form $X = A^{(n+1)}$, where A is a nominalistic theorem of **T'**, and n the Gödel number of a proof of A in **T**'. Σ is clearly recursive: in order to decide whether a proposition Z lies in Σ, start by checking whether Z is a conjunction, and if so, whether its second conjunct A, which is uniquely determined by Z, is nominalistic; we can then decide whether or not Z is of the form $Z = A^{(n+1)}$ where n must clearly be smaller than the (syntactic) length of Z; if $Z = A^{(n+1)}$, then n is uniquely determined by Z, for different values of n would give different lengths to $Z = A^{(n+1)}$; we finally have to check whether n is the Gödel number of a proof of A in **T'**. If and only if Z passes all these tests do we have $Z \in \Sigma$. If $Z = A^{(n+1)} \in \Sigma$, then $\mathbf{T}' \vdash A$, hence $\mathbf{T}' \vdash A^{(n+1)} = Z$. This holds for any Z in Σ; so $\mathbf{T}' \vdash \Sigma$.

G is defined as follows: the vocabulary of **G** is nominalistic, i.e. it consists of all the primitives of **T**' with the exception of $\in$, of s and of all abstract names. Thus the vocabulary of **T'** contains that of **G.** The recursive set Σ will be the class of all proper axioms of **G**. Since **T'** $\vdash \Sigma$, **G** is a subtheory of **T'**. Moreover, **G** yields all the nominalistic consequences of **T'**; for if **T'** $\vdash$ H where H is nominalistic, then let m be the Gödel number of any proof of H in **T'**; thus $H^{(m+1)} \in \Sigma$, so that, by definition of **G,** **G** $\vdash H^{(m+1)} \vdash$ H; hence **G** $\vdash$ H.

One must nonetheless agree with Field that such a **G** is a totally 'unattractive' theory, for its axioms are nothing but a sequence of 'powers' of its most important theorems. Still, **G** has the merit of resting on the well-defined concept of a recursively axiomatizable theory. **G** also clearly satisfies conditions (1) and (2) above. As for (3), note that mathematics has certainly proved useful in that **T'**, which enabled us to define the set Σ of the axioms of **G,** is a highly *mathematicized* theory. As defined above, **T'** includes $\mathbf{S_V}$, and hence Zermelo-Fraenkel set theory, thereby enabling us to develop all the known mathematical methods used in physics. But by furthermore demanding that nominalistic theories be, in his own words, 'reasonably attractive', Field exposes himself to an open-ended challenge; for every time a hypothesis such as **T'** is put forward, Field has to construct some 'attractive' nominalist alternative **G** satisfying (1)–(3). Attractiveness remains, if not a subjective, then at any rate a highly intuitive notion. It cannot be denied that some intuitive concepts are unavoidable in the sciences; but their number and significance should not be unnecessarily increased. In the present case, the notion of attractiveness was introduced in order to exclude intuitively unacceptable *nominalistic* alternatives to **T'**; but this problem would never have arisen, were it not for Field's positivist decision to restrict meaningful physics to being only about non-abstract entities.

Field seriously tried to carry out his programme only for the classical case of Newtonian gravitation, but Paul Chihara has convincingly demonstrated why this attempt can be taken to have failed. Field proposed a second-order nominalistic theory **Q**, for which he proved a *semantic* conservativeness theorem: he showed that if **M**$\cup$**Q** $\vDash$ g, then **Q** $\vDash$ g, where $\vDash$ stands for the semantic relation of logical consequence. But since second-order logic is incomplete, we cannot conclude that **Q** $\vdash$ g; and this, no matter how large a chunk of second order logic we formalize. Therefore, supposing that **M**$\cup$**Q** $\vdash$ g, then, by soundness, we obtain **M**$\cup$**Q** $\vDash$ g; semantic conservativeness yields **Q** $\vDash$ g; but we can no longer go over to the syntactical relation **Q** $\vdash$ g. And as perspicuously explained by Chihara, Shapiro proved that this last step breaks down for at least one nominalistic sentence g (Chihara 1990, Chapter 8). Note that qua nominalist, Field is anyway not entitled to fall back on semantic notions based on Platonist considerations of *whole classes* of models. Finally, second-order logic involves quantification over predicate variables. From an ontological viewpoint, it is difficult to see why such high-level logical operations should be more respectable than quantification over sets, even though all the extensions of the quantified predicates be singletons. Field admittedly outlined a first-order alternative to his theory **Q**; but Shapiro showed that the representation theorem might then fail, thus contravening desideratum (3).

Let me—somewhat unexpectedly—add that my major objection to Field's approach is not based on the technical hiccoughs mentioned above. These merely add fuel to my criticisms. By definition, nominalism is interested only in the non-mathematical consequences g of a given scientific system; and its chief concern is to find some suitable axiomatization of such g's without invoking any 'abstract' postulates. As a last resort, the nominalist can always fall back on a Craigean position which shows nominalism to be in principle feasible. As for 'attractiveness', it can be regarded as an expendable kind of luxury. Note furthermore that the Craigean theory **G** cannot be cogently criticized for having an infinite number of postulates: apart from being considered adhoc by eminent mathematicians like Poincaré and Borel, (ZFC) possesses infinitely many axioms; which did not prevent it from becoming the basis of all standard mathematics. A realist of my ilk would nevertheless maintain that anybody prepared to countenance either second-order logic or Craig's unnatural methods might just as well accept the *truths* of (ZFC) as a lesser evil. To which the nominalist might reply that according to *his* sense of reality, abstract objects simply cannot exist; for there is no space—of any description—for them to inhabit.

This sterile controversy illustrates, once again, the kind of *transcendent* criticism which was condemned by Popper: it consists of merely pitting one position against another incompatible with it. Such a debate can be settled by an appeal neither to logic nor to any empirical results. As explained above, Popper proposed another mode of criticism which he called *transcendental* as opposed to *transcendent:* it examines how well, or how badly, each philosophical system—taken separately—accounts for what is generally regarded as successful scientific practice (Popper 1979, Chapter 4, §9). In the same spirit, Lakatos had used the actual history of science in order to appraise various competing methodologies (Lakatos 1978, Chapter 2).

Applying the transcendental method, I now propose to demonstrate that some of the most important *scientific theses are irreducibly abstract* in that they inevitably refer to mathematical entities. Their actual derivation moreover depends on strong mathematical assumptions. Hence some of the most physically significant corollaries of a theory like **T'** are *non-nominalistic*, so that the whole rationale behind Field's project is called into doubt. The following examples will make the point. Scientists generally like to examine the conditions under which their fundamental hypotheses are deterministic. In his masterly review of 'Science without Numbers', Malament pointed out that deterministic theses cannot be rendered nominalistically (Malament 1982). Malament did not go into any details; but he might well have had the following argument in mind. The Newtonians claimed that, given suitable initial conditions, their theory would uniquely determine the motion of any set of particles. For their part, modern Relativists try to describe the Cauchy surfaces which fix the time-evolution of whole dynamical systems. It has finally been pointed out that despite being indeterministic with regard to individual events, Quantum Mechanics singles out a state function ψ which uniquely determines the expectation values of all observables. As already remarked, these claims assume the form $(\exists! \mathrm{f})[\mathbf{F}(\mathrm{f}) \wedge \mathbf{E}(\mathrm{f}) \wedge \mathbf{B}(\mathrm{f})]$; where we are often unable to solve the equation $\mathbf{E}(\mathrm{f})$ explicitly; i.e. f cannot be written down in terms of known functions; so we have to settle for an existence statement of the type just described. Being obviously mathematical, such a theorem cannot be nominalized; for: $\vdash \mathbf{F}(\mathrm{f}) \rightarrow \mathrm{s}(\mathrm{f}) \rightarrow \neg \mathrm{N}(\mathrm{f})$; so that any attempt to nominalize $(\exists \mathrm{f})[\mathbf{F}(\mathrm{f}) \wedge \mathbf{E}(\mathrm{f}) \wedge \mathbf{B}(\mathrm{f})]$ would lead to the contradictory formula $(\exists \mathrm{f})[\mathrm{N}(\mathrm{f}) \wedge \mathrm{s}(\mathrm{f}) \wedge \ldots]$. Moreover, in proving $(\exists! \mathrm{f})[\mathbf{F}(\mathrm{f}) \wedge \mathbf{E}(\mathrm{f}) \wedge \mathbf{B}(\mathrm{f})]$, we typically invoke the following interdependent properties of the continuum **R:** the existence of least upper (and of greatest lower) bounds, the compactness of closed bounded sets, the Bolzano-Weierstrass property, and so forth. All these results ultimately flow from the definition of the reals in terms either of equivalence classes of Cauchy sequences or of Dedekind cuts, that is: of infinite transitive sets of fractions. In other words, we assume the axiom of Infinity in order to determine ω; and we then make use of the power set axiom in order to define the real numbers. Not even the philosophical doubts of a Poincaré or of a Brouwer regarding all infinitistic methods prevented these great mathematicians from working within the classical tradition. Hilbert was therefore right about Cantor's paradise being very tempting. Two further examples underline the essential role which another synthetic mathematical principle, namely the Axiom of Choice, plays within science. In Quantum Mechanics, the states of a physical system are described by elements of a Hilbert space H. Qua linear space, H is taken to possess at least one base; that is, a maximal linearly independent subset B. The standard way of establishing the existence of B is via Zorn's Lemma which is known to be equivalent to (AC): starting from any linearly independent subset A of H, one defines the class $\mathbf{W}$ of all linearly independent sets X such that $A \subseteq X \subseteq H$. Using

Zorn's Lemma, one concludes that **W** possesses at least one maximal element *B*, which turns out to be a base of *H*. Even when defining an orthonormal basis of *H* by means of the Gram-Schmidt method, one needs *to choose*, at each stage, some vector which is linearly independent of the ones previously constructed.

Let me turn to my last example. It is often important to know that all the bases of a given *linear* space have the same cardinality. Again, the usual method of establishing this general result is by means of the equation $q^2 = q$, which is taken to hold for all infinite cardinals q; where this equation is again known to be equivalent to the Axiom of Choice (see Jacobson 1953, Chapter 9; for more details about the heuristic role of mathematics in the physical sciences, see above, 4.4 and 4.5).

Since they involve powerful mathematical principles and are moreover irreducibly abstract, the above results cannot be recuperated by the nominalist who has in effect renounced mathematics both as a language and as a theory. Though more broad-minded than *observationalist empiricism*, nominalism cannot but impoverish the *physical* content of scientific hypotheses. Some form of Platonistic realism is thus unavoidable; for having found that all the needed 'abstract' entities cannot be imbedded in the physical world, we must adopt a Platonist attitude with respect to those mathematical postulates which are implicated in strongly confirmed theories. It seems to me that our attitude should be a dual one. As far as the empirical basis is concerned, we ought to follow a Quinean 'desert landscape' policy: only phenomenologically reduced propositions—descriptions of our mental states without the use of any abstract predicates—should qualify as basic statements. But once a scientific system has been corroborated, it ought, provisionally, to be accepted as true or as possessing a high degree of verisimilitude; where 'verisimilitude' should be treated as an intuitive concept. Not only are all the consequences of the system to be taken seriously; its overall structure must also be regarded as having realist import.

These ontological issues will be examined in greater detail in the following, and last, chapter.

6

Formalism, Intentionality, or Platonism?

6.1 The Problem of Mathematical Ontology

In the previous chapter, we saw that in his philosophy both of logic and of mathematics, Russell started out by being a Platonist. He subsequently regarded these two disciplines not as having any specific content, but as describing the most general forms of thought. He was then driven to despair by the realisation that some of his basic assumptions, namely the axioms of infinity, of choice, and of reducibility did not have a strictly logical character. So he finally looked upon his system as an inductive hypothesis to be treated on a par with astronomy. Thus he had been, in turn, a Platonist, a logicist and an empiricist.

We also noted that in his long debate with Russell, Poincaré adopted a Kantian or rather a quasi-Kantian attitude. As is well known, Kant held logic to be a completed discipline which had been brought to a close by its inventor, Aristotle. As for mathematics, it was supposed to construct all of its concepts in the pure forms of sensibility: space and time; the former was taken to found geometry, and the latter, arithmetic. Having rejected the idea that we possess an apriori spatial intuition, Poincaré took the geometrical postulates to be conventions or implicit definitions of a system of primitive concepts and relations. Also unlike Kant, he held not only arithmetic but also logic to be based on the intuition of time: in the same way that mathematical induction is founded on the notion of temporal succession, logic rests on operations of classification conceived as processes carried out over finite time-intervals. Only if they prove to be immutable, or at any rate ultimately stable, do classifications legitimately define their intended objects.

In Chapter 1, we examined the way in which phenomenology enables us to circumscribe the empirical basis of the sciences. In the present chapter, we shall analyse the sense in which Husserl's concept of intentionality might help towards *reconstructing* Poincaré's views about logic and about mathematical ontology; for as repeatedly remarked above, Poincaré did not weld his ideas into one coherent system, so that a rational reconstruction of his various insights is imperative.

Let us start by describing, in very general terms, Husserl's problem-situation and his programme.

6.2 Husserl's Progamme

When considering the development of modern scientific thought, one is struck by the fact two philosophers of such different tempers of mind as Émile Meyerson and Edmund Husserl agreed on what they saw as the major flaw running through the whole of science. Instead of trying to give a general description of this flaw, let me follow Husserl's own method by tracing the origins of this *Problematik* back into Cartesian thought (Meyerson 1921, Book 4, Chapter 5; Husserl 1969a, p. 11).

It is well known that Descartes's contribution to philosophy is a dual one; dual in many distinct but related senses. On the one hand, Descartes fathered an ambitious and ultimately successful super-programme, namely the realist-physicalistic paradigm whose grand aim was to reduce physics to geometry. Geometry itself had been arithmetized and was taken to have thereby lost its qualitative aspects. Thus the heuristic of this programme was the science of continuous magnitude, which Newton and Leibniz later enriched by their discovery of the Differential Calculus. Let us call this whole strain of Descartes's thought the objectivist component of his programme. On the other hand, the certainty of all this objective knowledge was guaranteed by the subjective self-evidence of the Cogito. Through carriying out a transcendental reduction, in other words through abstaining, by means of a so-called phenomenological *époché*, from all assumptions concerning the existence of objects transcending the ego, the latter turns in on itself and thereby attains the certainty of the Cogito. According to Descartes, it is easier to know the contents of one's own mind than the nature of the external world. The ego posseses a clearer and more certain knowledge of itself—or rather of its own phenomenologically purified contents—than of anything else. Here we have the second strain of Descartes's philosophy, which will henceforth be referred to as its idealist or transcendental component. Descartes did not intend these two aspects of his system to remain disjoint; for God was supposed to mediate between them. More precisely, after establishing God's existence, Descartes went on to show that what happened clearly and distinctly in the mind faithfully mirrored what occurred in the physical component of reality (Descartes 1986,

Meditations 3 and 4). Thus the certainty obtaining in the transcendental sphere would flood into our knowledge of a mind-independent reality. The thesis of God's intervention in the world was therefore far from being an idle addendum to Cartesianism. But even Descartes's contemporaries, among them Pascal, realised that the Cartesian arguments for God's existence were invalid, and this not only because the ontological proof treats existence as an ordinary predicate, which was later shown by Kant to be inadmissible, but also because Descartes had taken an unwarranted leap in effecting a transition first from his own ego to the mind of the Divine Creator, and then from the latter to the physical component of his system (Kant 1787, Second Book, Third Section, 4). In modern terms, Descartes had used a hugely ampliative and hence invalid logic. There was moreover a hidden circularity in the Cartesian argument: Descartes conceded that God could have decreed different eternal truths, hence a logic different from the one in common use; yet these same truths had to be presupposed in the course of deriving God's existence from the certainty of the Cogito. Descartes had thus merely shown that if the Almighty existed and decreed *one particular* logic, then this logic could be used in order to establish His existence; which is a rather uninteresting result. It is well known that this circularity was noticed first by Leibniz and later seized upon by Husserl: if the phenomenology founded by Descartes was to remain self-sufficient, then it ought to constitute Logic from the certainty of the Cogito. We shall subsequently have to return to this point when dealing more directly with Husserl's programme. For the time being, let us restrict ourselves to examining the consequences which flowed from the invalidation of Descartes's proof of God's existence. Since the two components of Cartesian philosophy were now perceived to be essentially disjoint, each tried to go its own way, that is: independently of the other. This divided house was to have undesirable consequences for the progress of science. There was first the unwelcome dualism of mind and matter. There was secondly a divide between the absolute certainty founded on introspection, and a new heuristic based on the scientific use of powerful mathematical methods. A slightly different way of expressing the same thing is to say that predictive power was divorced from intelligibility, or that depth and certainty henceforth became irreconcilable ideals (Watkins 1984, Part 2, Chapter 4). We should furthermore note that all these cleavages were brought about by one fundamental schism within the Cartesian heritage.

There would have been less of a problem—or a mystery rather than a problem—had the schism been complete; but in at least one area, the objective and the subjective spheres seemed bound to overlap, namely in the observational domain. We have already seen that whilst observing, we appear to be relying on sense-data in order either to test some of our conjectures or to 'induce' laws from facts. Given this apparently inevitable point of contact, it was natural for philosophers to let one of the components of the Cartesian system either annex or at least override the other; or, alternatively, to show that there was after all an absolute divorce between the two. There followed not only various attempts at physicalist and idealist reductions, but also Neurath's and Popper's well-known theses that basic statements have to do with social conventions rather than with any observers' subjective experiences (Popper 1959, Chapter 5).

Let us now examine what is probably the most comprehensive and coherent attempt to appropriate the Cartesian heritage from a transcendental-subjectivist viewpoint, namely Husserl's phenomenology. By showing why and how Husserl ultimately failed, I hope to vindicate the view that all radical idealist positions must finally break down.

Having gone back to the basic Cartesian insight expressed in the Cogito, Husserl criticised Descartes for not being radical enough, in other words for not remaining faithful to his own *époché*; for Descartes had identified the ego with the human soul and hence with an objectively existing entity. If we are to believe Husserl, Descartes had forgotten that every object—including the soul—should be transcendentally constituted by certain acts of consciousness (Husserl 1969b, Meditation I). We have also remarked that despite his agnosticism about all 'eternal truths', Descartes had not only taken traditional logic for granted but also that he had beefed it up in order to obtain a proof of God's existence; where the latter was in turn supposed to guarantee the reality of a mind-independent world.

In Husserl's view, Kant fares worse than Descartes; for Kantianism actually meant the final abandonment of Cartesian radicalism. Despite his epistemological idealism, Kant remained a metaphysical realist in that he postulated a transcendent realm of things-in-themselves. He maintained that in order for there to be any appearances at all, 'something' must appear, and he called this 'something' a noumenon. To Husserl, this pseudo-argument must have sounded like a *naïveté* born out of Kant's ignorance as to the true nature of intentionality. According to Kant, experience comprises a passive moment through which things are intuitively given to us in time or in space-time; and a spontaneous or active moment through which the understanding processes sense-data by imposing on them an order ultimately derived from the unity of the transcendental ego. If we are to believe Husserl, Kant's thesis reduces to the view that the passive aspect of experience derives from a thing-in-itself totally independent of the ego. In

Erfahrung und Urteil, Husserl himself acknowledged the passive character of pre-predicative experience. He was however prepared to admit only a relative kind of passivity; more precisely: a semi-conscious activity which is less overtly voluntarist than any genuinely logical operation. Except for transcendental subjectivity which alone enjoys absolute existence, every Husserlian entity is thus created by the ego's constitutive acts (Husserl 1939, p. 16).

To repeat, Husserl held the postulate of a mind-independent noumenal world to be neither true nor false but meaningless. By its very nature, consciousness is intentional in that it is always a consciousness of something. Consciousness is thus directed at an object which can be either ideal or else real, that is: spatio-temporal. A logical connective, a proposition or a mathematical entity provides a good example of an ideal or categorial entity. It must therefore be possible to show that all objects are constituted by the ego's activity, from which they ought nonetheless to be sharply distinguished: since an intentional entity is that *at which* some mental activity is *directed*, it should never be confused with the latter. According to both Brentano and Husserl, the phenomenalisms of Berkeley, Mach, and Mill are untenably psychologistic; for these authors identify the intentional object with certain moments of the noeses—of the actual *psychological* processes. (See Mach 1922, Chapter 1). We should not however be misled by Husserl's protestations into believing that he was, after all, a realist. His so-called anti-psychologism is in effect a most radical form of idealist subjectivism.

As for the intentionality thesis, it entails that in perception we observe, not our own sensations, but spatio-temporal things given through their perceived properties and various relations between them. We can of course make the perceiving activity itself into a new object, thus turning it into an entity of a higher order—we go one type higher by directing our mental operations towards a fresh theme; for example: towards our *perception* of physical things, whereby this perception becomes distinct from the new activity directed at it. Note that it is precisely the task of the phenomenological reduction to raise us to a new type of consciousness (see Chapter 1 above).

Thus every object is immanent in one sense and transcendent in another. It is immanent in that it is generated by the subjective activity which targets it; and it is transcendent in that it forms no real part of such ego-logical activity. A Husserlian physical object is transcendent in yet another sense: it inadequately presents itself through the potentially infinite sequence of all its facets or *Abschattungen*, where such a sequence must inevitably remain incomplete (Husserl 1928, pp. 138, 143). In what follows, we shall not need this extra notion of transcendence, which was mentioned only in order to pre-empt confusion. To repeat, no genuine object is transcendent in the absolute sense of existing independently of the activity of all egos; though instinctively postulated by the 'naive' scientific attitude, such absolute transcendence is incoherent.

6.3 The Phenomenological Constitution of Logic and Mathematics

In the light of his intentionality thesis, Husserl levelled another criticism at Kant, a criticism which seems to me fully justified. As already mentioned, Kant had taken Aristotle's *Organon* for granted, thereby forgetting Descartes's basic insight that logic itself constitutes a problem rather than a ready-made solution to other problems; for God could have decreed different 'eternal truths'. Following his 'fundamental clue', Kant extracted an allegedly complete list of the Categories from Aristotle's twelve forms of judgment; whereas he ought to have previously derived logic from the unity of the transcendental ego. Given his radical thesis about the unknowability of the noumena, it was anyway impossible for him to perform such a task.

Similarly to Poincaré but by a somewhat different route, Husserl came to the conclusion that logic should be phenomenologically constituted—that it ought to be constructed by means of successive mental acts carried out in subjective time. It seems to me not only that Husserl's criticism of Kant is justified, but also that his intentional approach to all logical questions constitutes his major contribution to contemporary philosophy. Intentionality moreover provides a rationale for Poincaré's constructivist *yet* anti-psychologistic conception of the foundations of mathematics.

We have already said that most twentieth-century philosophers shy away from what they regard as the metaphysical extravagancies of Platonism. They hold that we have a valid reason for believing in the existence of a transcendent object only if the latter can act on us causally, that is: only if the object reveals its presence in some 'direct' way. This 'principle' is of course nothing but an empiricist prejudice, which is in effect undermined by the following observation: since matter and energy act on us by virtue of their *structure*, that is, since their effect on us depends on their *form* as much as on their *quantity*, nothing entitles us to deny causal efficacy to such—Platonic—Forms. Be that as it may, neither the logical constants nor any mathematical entities seem to fulfill this condition of causal efficacy. Like Husserl and Popper, we should naturally like to be able to say, on the one hand, that the logical connectives did not exist prior to our mental operations and hence to the emergence of consciousness in the evolutionary process; and, on the other, that once created, these categorial entities acquire an objective status and thereby become accessible to all thinking beings (Popper 1972, Chapter 4). As explained in the last chapter, Poincaré held similar views about the nature of mathematical objects.

Intentionality thus promises to yield the best solution to the foundational problems of both logic and mathematics. To repeat, on the one hand, all intentional objects are created *(erzeugt)* and constituted *(konstituiert)* by mental operations which must consequently precede them; on the other hand, these objects form no real part of the conscious activity or noesis; for they are merely generated by this noesis which simultaneously creates and targets them; the two processes of creating and of intending such entities occur at the same point of subjective time. According to Husserl, we have to do with objects which are abstract and yet non-psychological in the sense of constituting no real moments of any mental activity. We are thus entitled to claim that once created, these abstract entities become a sort of public property whose existence is on a par with that of ordinary things. The major difference between these two ontological categories is that the abstracta are, if not timeless, then at least supra- or omni-temporal. Husserl's thesis of the constitution—in subjective time—of atemporal objects is thus entirely coherent.

According to both Poincaré and Popper, the human mind creates not only *propositions* qua contents of well-formed formulas but also mathematical *objects* like the natural and real numbers. Once constituted, these entities acquire a kind of autonomy through exhibiting hitherto unsuspected properties; for the statements which define them have logical consequences of which we were initially unaware. For example, after formulating Peano's axioms, we realise, by means of complex proofs, that there exist infinitely many primes and that every positive integer can be uniquely expressed as a product of these irreducible numbers. Popper and Poincaré thus share Husserl's conceptualist views; but it was Husserl who, in *Erfahrung und Urteil*, did the hard work necessary for extracting logic from pre-predicative or pre-linguistic experience.

Its awkward style apart, *Erfahrung und Urteil* is a masterpiece in which Husserl largely succeeded in achieving the task he set himself, namely that of carrying out a phenomenological reconstruction of logic. Let me however start by mentioning one attempt in which he could not but fail. He never managed satisfactorily to define the *copula*, the basic relation between a substratum S and one of its properrties p, in terms of more basic phenomenological notions. He described this relation as some kind of overlap or of partial identity between S and p. It seems to me that such an explanation is bound to involve, at best, an unwanted circularity and, at worst, a category-mistake. A sliding-over-into, an overlap or a partial identity can occur only between entities of the same logical type; which is certainly not the case of S and p. Let us recall that even for the prelinguistic domain, Frege proposed a now well-known ontological distinction between any *saturated* object S and the predicate p which is always *in need of completion.* Husserl might of course have been trying to describe the copula in an informal manner rather than to explicate in any rigorous terms; for being presupposed by the very act of perception, the copula cannot be reduced to more fundamental perceptual experiences. In Kantian language, we could say that since experience is made possible by the substance-and-accident category, the latter cannot be induced *a posteriori* from any psychological facts. (No presupposition as to the conservation of substance need naturally be made.)

Let us now turn to the logical constants: $\neg$, $\wedge$, $\vee$, $\forall$, and so forth, whose phenomenological origins were successfully explained in terms of our receptive or pre-predicative experience. Husserl founded the construction of these connectives on the temporal dimension inherent in the perception of all concrete objects (Husserl 1939, Chapter 1, p. 21). It is this temporal component which enables us to anticipate, in a more or less fallible manner, the successive facets or *Abschattungen* of a concrete entity. More precisely, perception never consists of an instantaneous grasp of a substatum S considered in isolation from the rest of the world; a whole background is given simultaneously with the object S, which is thus apprehended against an external as well as an internal horizon. The external horizon adumbrates a system of possible relations between S—qua focus of attention—and certain objects of a background which is tacitly registered rather than explicitly thematised. The internal horizon pre-indicates a whole sequence of properties of S, thus giving rise to certain expectations regarding these possible predicates. Whenever these expectations are verified, they fill out our incomplete image of S. It may however happen that an expectation is frustrated, disappointed or thwarted. Note that we must avoid using the word 'refuted' which not only involves a vicious circle but also fails accurately to describe our pre-predicative experience. Our 'disappointment' constitutes the foundation of a categorial constant, namely the logical negation operator $\neg$. The latter is articulated at a level higher than the purely receptive one: it belongs to the genuinely predicative or linguistic domain which nonetheless presupposes the receptive stage. Husserl gives the following example: I expect the hitherto unseen side of some object to be red; on inspection, it turns out to be green and I consequently experience a feeling of disappointment. My feeling is caused, not by the presence of the green colour, but by the *absence* of the anticipated redness. It is this experience of absence, of our expectation being somehow thwarted, which is subsequently codified, at the logico-linguistic level, by the symbol of negation. Let us moreover note the fundamental role played by the intuition of time in the constitution of this first logical connective.

Similar considerations apply to the conjunction $\wedge$. The substratum S unfolds into its successive determinations: a, b, c, and so on. These provide the pre-predicative foundation of the conjunction functor, which initially appears within the context of a temporal sequence of predications: S is a&b&c&.... It is only during the predicative phase that through a kind of transfer, we constitute $\wedge$ as an operation conjoining one proposition to another. We no longer go from S directly to the series of its properties: a, b, c,... ; every time a property is grasped, we go back to S as to the substratum possessing each determination in turn. Having reached the properly linguistic level, we construe the conjunction as a connective linking one proposition to another: (S is a) $\wedge$ (S is b) $\wedge$ (S is c) $\wedge$ .... As for the universal quantification $\forall$, it can be regarded as indicating an indefinite—or potentially infinite—conjunction of predicates or statements.

According to Husserl, we are now on a plane from which we can rise to an even higher level of mental activity: we thematise the meanings of such composite linguistic entities as terms, sentences, and well-formed formulas. Through substantification, we constitute a *proposition* as the sense or meaning of a statement taken as a whole. In their turn, such meanings assume the roles of substrata in propositions of a higher order. Corresponding to every sentence q, we finally create an entity—also denoted by q—which is based on a system of objects and on their properties and the relations they bear to one another: this is the *state-of-affairs* corresponding to the sentence q. Let us again underline the fact that in the subjective constitution of the connectives, a central role is played both by the intuition of time and by the protentional awareness of a horizon given alongside all intentional objects. This does not mean that qua noemata, the logical constants have any temporal connotations; but that far from being instantaneous mental acts, the creative noeses giving rise to these categorial entities are spread out over finite time-intervals. All ideal objects are therefore constituted in time, but the latter does not form part of their determinations: such entities possess no temporal properties. For example, qua intentional object, a proposition is created in time, but it is not intended as something which has any spatio-temporal location; for although the *same* proposition can be envisaged at two distinct instants, it would be impossible for us to experience the same event at two different points of time.

Husserl held a proposition q, conceived as the meaning of a statement, to be omni-temporal rather than a-temporal. In other words, instead of regarding q as meaningful or true in a *timeless* way, Husserl held it to possess both a sense and a truth-value *at all points of time*. This position however seems to me to give rise to insurmountable metaphysical difficulties. For example, let q describe, through mention of precise dates, a sequence of events which unfold after some instant t_0. If we adopt Husserl's viewpoint and if q happens to be true, then we must regard q as holding good at *all* instants and hence before t_0. Thus a description of all the initial conditions obtaining at t_0—the set of all the propositions true before t_0, to which q naturally belongs—determines the course of history after t_0. In itself, determinism is admittedly a coherent thesis; but we have here a perplexing variant of *logical* determinism which seems to follow from a mere convention. Such an 'argument' in support of determinism would however lose all persuasive power if we take q to be a-temporally rather than eternally or omni-temporally true; for q would then cease to express an initial condition at t_0.

6.4 The Ultimate Failure of Husserl's Programme

Provided we replace the condition of the omni-temporality *(Allzeitlichkeit)* of all propositions by that of their a-temporality *(Zeitlosigkeit)*, we can justifiably maintain that Husserl succeeded in founding first-order logic on pre-predicative experience. Husserl's success is not however due solely to his phenomenological analyses, profound though they are; it also follows from the fact that first-order logic satisfies a requirement laid down by Husserl himself in *Ideas:* in order to determine a field of categorial entities and more particularly the domain of all mathematical objects, an axiomatic system must prove self-sufficient in that every question formulable in the system is decidable within the system itself. The latter should, in other words, be complete. Husserl ought to have mentioned another condition connected with that of completeness, namely the requirement of categoricity: all formal systems must determine their domain to within an isomorphism.

In what follows, we shall restrict ourselves to the completeness condition. In 1931 Gödel established the completeness of first-order logic, so that the semantic notion of logical consequence was shown to be adequately captured by that of syntactical derivability. Hence all analytic truths can be formally deduced from first-order logical axioms. It seems to me that thanks to this technical result, human consciousness can justifiably be credited with having created first-order logic. Should we hold the latter—as Quine certainly does—to be the only system worthy of the name 'logic', then we can conclude that Husserl achieved the aim he set himself in *Erfahrung und Urteil*; but as already mentioned, he had as a matter of course also assumed that all formal systems are complete. Like Poincaré before him, Husserl had intuitively taken the set of all mathematical truths to be recursively enumerable and hence axiomatisable. But as Gödel was about to prove, every axiomatic system containing one of the most elementary mathematical theories, namely arithmetic, is necessarily incomplete. If Husserl and Poincaré can be excused for not having either known or else correctly assessed the significance of Gödel's work, it is astonishing to find Popper and Tragesser still maintaining that mathematical objects are created by the human mind (Popper 1972, Chapter 4; Tragesser 1984, Chapter 5). Given the completeness of logic and the formal undecidability of mathematics, the maximum we could reasonably lay claim to is our ability to construct semantically adequate formal systems. It might even be maintained that such systems determine, if not a unique domain, then at least a whole class of models. Model theory is however based on set theory, which is usually taken to be an interpreted rather than a formal system; but set theory is in turn incomplete: it does not determine its domain and must be regarded either as a formal system or else as a hypothesis which does not construct but merely characterizes its objects. Building a set-theoretic model of a formal system comes down, in effect, to mapping one theory into another. This is why most mathematicians adopt one of two seemingly extremist positions: Platonism, or else out-and-out formalism. It can be maintained that while practising Platonism in secret, most formalists pretend to be indifferent to foundational problems. It is therefore no accident that unlike the *pure mathematician* Poincaré, the *mathematical logician* Gödel adopted an explicitly Platonistic rather than a Husserlian stance; for Husserl's conceptualism falls between two stools: it suffers from the disadvantages of formalism *and* of Platonism, while renouncing the positive aspects of *both* positions. Conceptualism subscribes to the existence of abstract entities like numbers and functions but it cannot attribute any meaning to questions concerning the truth-values of formally undecidable propositions. The latter must therefore be declared, by fiat, either true or else false; either they or their negations must consequently be adjoined to some axiom system in order to decide certain problems. Such a process cannot however be completed; so we never uniquely determine, let alone create, all the mathematical entities we need. It seems to me preferable to postulate the independent existence of such entities—together with some of their properties—and then draw consequences from these assumptions. In other words, we should opt for a fallibilist version of Platonistic realism which would make sense of the discussions and disagreements between practising mathematicians. If we are prepared to take philosophical risks by going beyond pure formalism, it is preferable to go the whole hog by adopting Platonism rather than stop half-way and accept Husserl's or even Poincaré's version of constructivism. Gödel speaks of a pregiven universe of sets and of our ability to grasp these mathematical entities, which is comparable to man's capacity for perceiving concrete objects (Gödel 1983a, p. 483). Thus neither Zermelo-Fraenkel nor Gödel-Bernays set-theory actually generates the universe of sets, though each system tries to single out certain pre-existing classes and describe their properties. Despite the formal undecidability both of the Axiom of Choice (AC) and of the Continuum Hypothesis (GCH), Gödel regarded the first proposition as true and the second as false; so that his position would be incomprehensible without strong Platonistic assumptions.

With respect to mathematics, even an empiricist like Quine effectively adopted a Platonistic attitude; for Quine rightly takes quantifications over set variables at face value. He holds that whether we are aware of it or not, such quantifications reveal our commitment to the existence of classes. He admittedly looks upon logic and mathematics as revisable systems: according to him, physical theories confront the verdict of experience en bloc; so that in case of refutation, we could decide to modify, and hence implicitly regard as refuted, not only the mathematical but also the logical postulates (Quine 1980c). It seems to me that by taking a fallibilist view of both logic and mathematics, Quine has in effect adopted a realist attitude with regard to these two disciplines. He is therefore a Platonist as well as a fallibilist. Thus the choice facing logicians and mathematicians is still the traditional one: either formalism or Platonism; or rather: Platonism during mathematical practice *and* formalism whenever the mathematician has to defend his position from a philosophical viewpoint, whence a fundamentally schizophrenic situation. This explains Poincaré's blatant double standards: he was both a philosophical conceptualist and a Platonist practitioner of mathematics. For example: he did not—and certyainly could not—effectively define all the real numbers mentioned in his work on classical analysis.

Like all the mathematicians of his time, he operated with numbers on the implicit assumption that they are pre-given to, and hence not created by the the mind's operations. Yet in his controversy with Russell, he insisted that all mathematical entities must be constructible in the pure intuition of time.

In conclusion, I hope to have demonstrated that even in the philosophy of mathematics where conceptualism seems to be the appropriate response, pressures towards a more overt form of realism are strongly felt.

6.5 Positivism and Idealism versus Poincaré's Structural Realism

Mach's brand of idealism exerted a strong influence not only on the young Einstein but also on Duhem and Poincaré, both of whom often cite him in their works, without however sharing his phenomenalist metaphysics. This is why we should say something about subjectivism in the philosophy of physics; and, more particularly, explain why, despite having often been envisaged, idealism ultimately broke down. To repeat, according to Husserl, every object presupposes a subjective activity through which the object is constituted. To speak of an entity as though it were totally independent of any transcendental ego is nonsensical; so that the objectivist realism professed by most natural scientists should be regarded as neither true nor false but meaningless. Given this meaning criterion, it looks as though no rational debate between realists and phenomenologists were possible.

Against Husserl, it could of course be maintained that meaning-criteria are language-dependent and hence relative; which he would of course deny. So the debate would again have to stop at this point. As repeatedly mentioned above, there is however an *immanent* method of criticising both Husserl and all the positivists of the Machian type; namely a method of attacking these idealist philosophies from within, that is: by showing them to be incapable of solving the very problems they set themselves. Husserl and Mach questioned neither the methods used, nor the success achieved by the natural sciences; they merely set out to account for this success in a metaphysics-free way. Husserl proposed to go back to the transcendental subjectivity through which all objects of ordinary experience and all physical notions and hypotheses are allegedly generated. Mach had similarly wanted to free the sciences from the metaphysical gunge which 'obscures' their true meaning. Like Berkeley, he wrongly took material bodies to be relatively stable complexes of sense-impressions which he called 'elements of sensation'. Following Brentano's lead, Husserl realised that while sense-impressions form an ever-changing flux within the stream of consciousness, the objects intended both by theoretical science and by everyday experience are constant entities *given through* this flux. In other words: science and commonsense need to refer to fixed objects like tables and chairs, electromagnetic waves and elementary particles, and so forth (Husserl 1969a, §9). We have already explained in what sense consciousness is intentional: through its *Erlebnisse*, it generates and simultaneously targets entities which, once created, are always available for the ego to recall. Thus Husserl's approach, which had anyway proved its mettle in Logic, was bound to prove vastly superior to Berkeley's and Mach's. Should we therefore succeed in undermining phenomenology as applied to the natural sciences, then we would *a fortiori* have considerably weakened Machian positivism (Mach 1933, Introduction).

Despite their differences, all the idealist positions described above share certain features which expose them to unanswerable criticisms. For example, they assume that all entities, no matter how they are defined, must be subjectively constituted and that the 'theoretical' entities posited by the natural sciences are idealisations of ordinary macroscopic objects. In the *Krisis* Husserl refers to the domain of everyday experience as our *Lebenswelt*; so that the meaning and validity of the most sophisticated scientific hypotheses are ultimately to be grounded in this Life World. This is why Husserl accused the natural scientists of wrongly regarding their hypotheses—which he takes to be mere methods of idealisation—as having ontological import. In the last analysis, Husserl's position was therefore not far removed from Machian instrumentalism.

Already at this stage, we can formulate a serious objection to Husserl's philosophy of science. Physics makes a clear distinction between the situations in which it idealises an 'appearance' and those in which it postulates real objects causally linked to that selfsame 'appearance'. Geometrical optics deals with rectilinear rays whose properties are known to conflict both with precise observation and with electromagnetic theory; so only the latter is realistically interpreted while the light rays are treated as mere idealisations. Another example is provided by the assumption of a continuous distribution of matter which Poincaré knew to be a convenient fiction. These cases are very different from those in which Mechanics postulates various sets of particles in order to *explain* our perception of macroscopic 'bodies'. These bodies are in effect intentional entities arising from the presence of the particles. Since the latter are taken to be real, they certainly do not 'idealise' any observed object in the sense in which a rectilinear beam idealises a wave process.

By not distinguishing between these two cases, Husserl fails to account for the success of the sciences which is largely predicated on such a distinction. According to him, all talk of hidden physical entities causing the perception of Lebenswelt objects is nonsensical; for, as often repeated, all entities ought to be subjectively constituted. This implies that all physical objects should, in the last analysis,be observable. This clearly question-begging argument has given rise to so many controversies that I feel obliged, once again, to explain my quasi-Cartesian position in some detail. By means of long causal chains, some events in the physical world act on other physical entities, namely on our brain processes. There exists a correlation between the latter and certain states which we describe as 'mental'. Such a parallelism, which need not be one-one, may strike us as something mysterious or as a totally irrational jump from one domain to another (see Meyerson 1907, Chapter 9). It would nonetheless be irrational to deny the existence of such correlations. If I do nothing but repeat Descartes, it is because independently of any ontological issues like those of monism and dualism, some Cartesian insights seem to me both relevant and definitive. Thus there exists between some physical processes on the one hand, and the Lebenswelt with its intentional entities on the other, a relation not of idealisation but of the kind of causality which I have called 'correlation'; where this term was chosen on account of its neutrality. Precisely for this reason, Brentano made a fundamental distinction between two sorts of psychology: purely descriptive on the one hand, and genetic or causal on the other (Brentano 1982, I). The former ought to remain autonomous and is in fact indiscernible from phenomenology; the latter must appeal to the natural sciences, and more particularly to physiology in order to explain the time-evolution of our conscious activity. All the sciences, including genetic psychology, presuppose phenomenology as their common empirical basis (see above, Chapter1).

Let us recall that Husserl's aim was to render the success of the sciences intelligible by ridding them of all allegedly nonsensical talk. But as just demonstrated, this success goes hand in hand with the 'nonsense' which Husserl was consequently unable to cut out. In this respect, the failure of Husserl resembles that of Mach, who set out to eliminate all unobservables from Newtonian theory while trying to keep the latter's empirical content intact. Through attempting to rid himself of the 'monstrous' concepts of absolute space and time, Mach however found himself landed with *Mach's problem:* from a positivist standpoint, the oblateness of the earth must be attributed solely to its rotation against the stars; so what would happen if the latter were to disappear one after the other? More precisely: what would happen under changing boundary conditions in which the total number of stars is made gradually to decrease? (Mach 1933, Chapter 2) According to Newton, nothing essential would be altered; for he held the earth's oblateness to be due to its rotation, not relatively to the stars, but in absolute space; the shape of the earth would consequently remain unaffected by any alteration of the initial conditions. For Mach however, with the disappearance of the last star, it would become meaningless to speak of the earth's movement; so that the oblateness should instantaneously vanish, thus causing a violation of the principle of continuity. Though acutely aware of this problem, Mach refused to solve it by resorting to what he regarded as 'metaphysical nonsense', for instance by postulating the earth's absolute motion. Nothing illustrates more vividly the positivists' suicidal attitude.

Let me now demonstrate that by attacking idealim, I am not 'flogging a dead horse'. Mach's influence on Einstein is well-known and we have already noted Mach's inspirational role in the genesis both of Duhem's and of Poincaré's philosophies of science. More recently, physicists have been perplexed by the structure of Quantum Theory whose wave-functions appear to require a subjective-probabilistic interpretation; also by certain propositions like the anthropic principle which seem to imply that the universe was created especially in order to permit the emergence, within it, of the phenomenon of consciousness. (Had the fundamental constants been different, then life and hence also consciousness would not have arisen or evolved; see Barrow and Tipler 1986, Chapter 8). Some thinkers have jumped to the conclusion that prior to the existence of consciousness, the physical world either did not exist or else was a totally structureless mass. They allege that the emergence of human thought created an order which the mind imposes—both retrospectively and henceforth—on a hitherto amorphous universe. Without being fully aware of it, some physicist-philosophers in effect revive Husserl's old theses; for remember that according to Husserlian idealism, consciousness starts by constituting our *Lebenswelt*, the world of everyday life. It then creates physical theories as a mere mantle of ideas which is thrown over the Lebenswelt and whose only function is to facilitate the prediction of events occurring in this Life World. We have now reached a point at which Poincaré's no-miracles argument can be used against Husserl; for if Husserl is right about our having constituted the world of physical objects, then why should consciousness have fabricated such unintelligible hypothess as Quantum Mechanics and Quantum Field Theory in order to account for prosaic everyday experience?

Furthermore, unless we be prepared to accept the occurrence of miracles or of whole series of coincidences, the frequent predictive successes of such counter-intuitive theories cannot, according to Poincaré, be ascribed to mere chance. The unexpected confirmation of high-level hypotheses can be explained only by assuming the existence of an independent reality whose structure is approximately modelled by our conjectures. This is, in a nutshell, Poincaré's structural realist thesis (see above, Chapter 2).

Husserl seems not to have noticed that the problems confronting him were the opposite of those faced by Descartes. The latter had to explain how our clear and distinct ideas, how certain notions intelligible to our minds, described a world independent of human thought. He concluded that such a correspondence between thought and reality was warranted by God's veracity. Having renounced all transcendent 'proofs', Husserl tried to account for this correspondence by holding objective reality to be a product of our mental activity. From this assumption it follows that being constituted by human consciousness, the external world will appear intelligible to us. Husserl however failed to take stock of the reversal of this problem-situation which had taken place between 1905 and 1925, a reversal which was ushered by Poincaré's seminal work on the foundations of geometry and mechanics. In *Ideas*, in *Formal and Transcendental Logic*, and even in *Experience and Judgment*, Husserl still refers to physical geometry as being necessarily Euclidean, to all objects as being exactly localised in absolute space and time and to mathematical theories as forming complete systems. It is true that in their gravitational theories, Poincaré and Einstein had gone back to the contiguous action held by Descartes to be the only intelligible form of physical causation. Poincaré and Schlick however realised that Euclidean geometry and the absolute time hypothesis were conventions to be adhered to only as long as they remained convenient; and during Husserl's lifetime, Einstein had shown the necessity of giving up both of these hypotheses. For their part, the Quantum Physicists abandoned the requirement of precise spatial localisation and introduced concepts which not only failed to possess any classical counterparts, but also proved incompatible with assumptions made about macroscopic objects. Let us, for example, consider the notions of indiscernibility and of non-separability. Unlike macroscopic bodies, quantum objects are not strictly individuated: two numerically distinct but otherwise identical quantum particles are not even in principle discernible; we are therefore not entitled to assume that permuting two such entities would give rise to a new situation which, though equivalent to the old one, is distinct from it. As for non-separability, it says that certain quantum systems are such that without reference to a whole set-up, we can meaningfully speak neither of the properties nor even of the separate existence of its parts; so that every partitioning into subsystems becomes illegitimate. Thanks to Bell's inequality and to Aspect's subsequent experiments, this consequence of Quantum Theory became testable. Let us note that Bell's inequality is deduced—totally independently of the quantum postulates—from minimal assumptions which seem underwritten by our most familiar Lebenswelt experience (see Redhead 1987, Chapter 4). Aspect's experiments have refuted these seemingly ungainsayable presuppositions while simultaneously confirming Quantum Mechanics. It should be pointed out that as opposed to a double confirmation, a double refutation was a priori possible. The predictions of Quantum Theory are admittedly incompatible with the minimalist classical assumptions which entail Bell's inequality; but both predictions could in principle have been refuted. There has therefore been a double blow: on the one hand, the falsification of a weak and 'self-evident' postulate; on the other, the corroboration of a counter-intuitive consequence of Quantum Mechanics. Thus, although Nature forbids all instantaneous actions-a-distance, there seem to be correlations at considerable distances between certain elementary particles. The latter form wholes which are indivisible in the sense of being *conceptually* inseparable. As explained by Michel Paty, this kind of non-separability was always inherent in Quantum Theory: some state-vectors, which depend on the coordinates of all the particles forming a system, cannot be mathematically expressed as two juxtaposed functions (Paty 1988, Chapter 6).

The counter-intuitive character of modern physics undermines the idealist-phenomenological thesis that all objects without exception are constituted by human consciousness; for why should the latter find some of these objects—which ought to be its own products—so utterly incomprehensible? As pointed out by Eddington, science conclusively shows that stubborn irreducible facts compel us reluctantly to construct theories which are totally unintelligible to the human mind (Eddington 1958, Chapters 1, 5). Thus we come up against obstacles which, being alien to our understanding, must perforce be ascribed to some mind-independent reality.

There is one final objection to idealism which shows the full relevance of Poincaré's no-miracles argument. Schopenhauer expressed this objection in the form of an antinomy to which all causal explanations are exposed. Like Husserl but in much better style, Schopenhauer assumes that there can be no object without some conscious subject; and less idealistically than Husserl, he held that that any attempt to derive the subject from the object, or *vice versa*, was bound to fail; for it was based on a misunderstanding of the nature of these two complementary poles. Against this view, Husserl maintained the primacy of the transcendental ego, which alone leads an independent existence while all other entities are intentionally constituted by its operations. However, these divergences between the two authors' views about the relationship between subject and object make no difference to the derivation of the idealist antinomy, which is as follows. In investigating Nature, we are inevitably driven to the conclusion that animal, and *a fortiori* human, consciousness arrived late on the evolutionary scene. Before this late arrival, there was a long inanimate stretch during which only lifeless objects existed. Yet according to idealism, the existence of every object entails that of a corresponding subject. We are thus led to the conclusion that this inanimate period popped into existence concurrently with the emergence of the first conscious being, who could nonetheless justifiably regard this same period as having preceded him. Schopenhauer took this paradox to be an inevitable consequence of the principle of sufficient reason. Qua transcendental idealist, he tried to wriggle out of this antinomy by maintaining that causal explanations apply exclusively to the world of appearances, which is the manifestation or objectification of a unique noumenon: the Will. The latter can be adequately apprehended only through a passive contemplation of Platonic Ideas not involving the Principle of Sufficient Reason, that is: not involving the principle responsible for the paradox. Such Platonic empathy is taken to be immune to the above antinomy, whereas the knowledge accessible to ordinary human understanding will forever remain paradoxical.

Husserl was too genuinely concerned about both logic and the natural sciences to accept such a let out. As explained above, he in any case regarded the noumenal thesis as unacceptably naive; for in it, we conceive of a thing-in-itself which is totally independent of any subjective constitution, and this he took to be a contradiction in terms. He admittedly distinguished between the subjective time-intervals during which all intentional objects are constituted and the objective time in which some of these objects are meant to be located.

Let us incidentally note that without any justification, he rejected solipsism and considered every other ego as a transcendent entity on a par with his own self (see Husserl 1939, p. 64). But the above antinomy is not thereby eliminated; for the whole of Nature, including all psycho-physical bodies and hence also all psychological egos, have now to be placed within one objective time-order. Each ego can carry out a phenomenological reduction similar to my own and then reconstruct an intentional world which proves to be isomorphic to mine. The paradox re-emerges in that there ought to be—within objective time—a point before which there was no life, no psychological egos, hence no transcendental egos and therefore nothing at all. Yet physical science tells us that before that point, there existed all kinds of inanimate objects. Worse still, the emergence of life might well have been improbable; so that a world with no life and hence no consciousness in it at all could have evolved. Husserl is therefore compelled to declare—*a priori*—that such a state-of-affairs is impossible; hence he cannot but ride rough-shod over the very sciences he set out to rescue from absurdity. He could of course have regarded transcendental subjectivity as being independent of all individual egos—as a kind of impersonal Hegelian Spirit through which all objects are constituted. Such a position would however differ only verbally from outright realism, for it would enable us to speak of objects pure and simple, without having repeatedly to refer to that unique Spirit.

Thus no genuine idealism, no matter how mitigated or improved, can be squared with evolutionism; and it is amazing that Mach should have proposed one of the first Darwinian accounts of perception. He put forward the view that under certain evolutionary pressures, animal species may develop a new capacity for discerning those colours which are connected with the food they urgently need (see Mach 1926). Evolution thus brings about the modification and even the emergence of certain kinds of sensation, that is: of certain mental states. Had Mach been asked about the physical entities which exert these evolutionary pressures, then he would have had to reassert that they also consist of elements of sensation present to the mind of the animal undergoing the evolutionary change. Such an answer involves a palpably absurd circularity; and it is no wonder that Mach never really addressed the corresponding question, which ought nonetheless to have been obvious to him. Fortunately, Mach's philosophical incoherence is also the reason why masterpieces like *The Science of Mechanics, The Conservation of Energy*, and *Knowledge and Error* can be appreciated without our having to take empirio-criticism seriously.

To sum up, if taken seriously, idealism is reduced to accepting the thesis that the emergence of consciousness brought about the existence either of physical reality as such, or else of the structure of a material world which had hitherto remained amorphous. The creation of matter through mind must moreover have been such as to enable us successfully to explain, by means of scientific hypotheses, present facts from initial conditions which preceded the emergence of consciousness, that is: from conditions obtaining at a time when the physical world was either non-existent or else devoid of any structure. Such a conspiracy theory not only constitutes the most absurd denial of Poincaré's no-miracles argument; but it also shows that idealism, which is often prepared to believe in such conspiracies, dies very hard.

6.6 Conclusion

Against Poincaré's and Husserl's constructivist conception of mathematical ontology, I have argued that the human mind may be credited with the creation of a *complete logic* together with the construction of essentially *incomplete theories*; the latter capture some, but certainly not all of the properties possessed by the fundamental mathematical objects. Thus neither the latter nor the sum total of their properties could have been created by human consciousness. It follows that no matter how prima facie incredible, fully-fledged Platonism, or else formalism, is the only way out of this dilemma. In as far as the mind can be said to have constructed *logic* and number *theory*—as distinct from the numbers themselves—Poincaré, Husserl, and Popper appear to be saying essentially the same thing. Against Popper, Husserl could legitimately have claimed that although we might be unaware of the derivability of certain propositions from some high-level theory, yet by determining a set of axioms together with a complete logic, we would implicitly have generated all the theorems of the theory in question. But to repeat: it seems to me that only the *theorems*, as distinct from the *entities* they refer to, are all that we can legitimately claim to have created.

When we come to the physical sciences, Husserl's and Mach's idealist programmes break down completely. We therefore have to accept the existence of a transcendent world together with Poincaré's structural-realist thesis with regard to it; also his no-miracles argument concerning all unified and highly corroborated—yet fallible—theories about such a world.

Appendixes

We have seen that the principle of the unity, or of the internal perfection of a theory is at the heart of Poincaré's methodology. Irrespective of whether or how this notion should be formally explicated, one has to guard against its being trivialised in the following manner. Start from two hypotheses A and B which have nothing in common; A could for instance be a psycho-analytic conjecture and B a low-level physical law. No system could therefore be less unified than the one consisting of the two postulates A and B. This system will henceforth be denoted by S. To the intuitively self-evident objection that all the strictly physical predictions of S are already entailed by B, a devil's advocate might answer that instead of {A,B}, the axiom set of S should be taken to consist of the singleton {H}, where $H \equiv (A \wedge (A \Rightarrow B))$. He would moreover correctly remark that in order to derive any empirical consequences from H, one needs to invoke, not A alone or $(A \Rightarrow B)$ alone, but both A and $(A \Rightarrow B)$. It now looks as though S were a highly integrated theory possessing a unique postulate H, both of whose conjuncts play an essential role in every scientific explanation. This counter-intuitive conclusion is made even worse by the fact that both A and B normally consist, not of atomic propositions, but of quantified sentences. For example, A and B might well be of the form: $A \equiv (\forall x)(\exists y)R(x,y)$, and $B \equiv (\forall x)P(x)$.

Taking the variables x, y, z, u, v to be distinct, it can be easily verified that:

$$\vdash H \Leftrightarrow [A \wedge (A \Rightarrow B)] \Leftrightarrow (\forall x)(\exists y)(\forall v)(\exists z)(\forall u)[R(x,y) \wedge (R(z,u) \Rightarrow P(v))].$$

Thus the prenex normal form of H looks like one unified formula. Yet we know that H is equivalent to $A \wedge B$, where A and B have disjoint domains of discourse. We should therefore like objectively to be able to claim, on the one hand that $\{A \wedge (A \Rightarrow B)\}$ is an *unnatural* way of axiomatising S and, on the other, that when *naturally* axiomatised in the form {A,B}, S is a *non*-unified system. In other words, Poincaré's unity criterion must be regarded as subjective as long as no formal definition of the naturalness of an axiomatisation has been given. More generally, a hypothesis can be declared either unified or else disconnected only *after* it has been *naturally* axiomatised in such a way that its postulates are *logically independent* of one another. We therefore need to explicate—as precisely as we can—the notions of logical independence and of natural axiomatisation. Note that the converse of these claims need not hold: an objective criterion of naturalness will not automatically yield an objective definition of unity, whence the need for a (third) appendix on the unity of scientific systems.

APPENDIX I

Logical Independence

[A] A set Δ of sentences is said to be logically independent if no element of Δ logically follows from the remaining members of Δ; i.e. if Not $[(\Delta - \{X\}) \vdash X]$ for all $X \in \Delta$.

[B] For any set Δ of formulas, $<\Delta>$ will denote the class of all logical consequences of Δ. Thus $\Delta \subseteq <\Delta>$; $<\Delta>$ is moreover closed under the relation of logical consequence; i.e. $<<\Delta>> = <\Delta>$.

[C] A theory Ω is any set of sentences closed under the relation of logical consequence; i.e. $<\Omega> = \Omega$. Ω will henceforth denote a first-order theory expressed in a denumerable language; and Ω_0 will be an arbitrary axiom system of Ω; e.g. $\Omega_0 = \Omega$. More generally: $\Omega_0 \subseteq <\Omega_0> = \Omega$.

[D] We propose to show that Ω possesses at least one logically independent set of axioms; i.e that there exists at least one logically independent subset Σ of Ω such that $<\Sigma> = \Omega$.

[E] Since the underlying language is denumerable, we can write:

(1) $\Omega_0 = \{A_0, A_1, A_2, \ldots\} = \{A_i : i \in \omega\}$; where, as usual, ω denotes the set of all natural numbers. Let us define the sequence of sentences $B_0, B_1, \ldots$ as follows:

(2) $B_0 \equiv A_0$; $B_{n+1} \equiv ((A_0 \wedge A_1 \wedge \ldots \wedge A_n) \Rightarrow A_{n+1})$. Put

(3) $\Sigma_0 = \{B_0, B_1, \ldots\} = \{B_i : i \in \omega\}$.

[F] To show that $\Sigma_0 \subseteq \Omega$.

Proof: We have the tautology $\vdash A_m \Rightarrow ((A_0 \wedge \ldots \wedge A_{m-1}) \Rightarrow A_m)$; i.e. $\vdash A_m \Rightarrow B_m$. But by definition of Ω_0 and of Ω: $A_m \in \Omega_0 \subseteq \Omega$. Thus: $A_m \in \Omega$. Hence $\Omega \vdash B_m$; from which $B_m \in \Omega$ follows, for Ω is closed under logical consequence. By (3): $\Sigma_o \in \Omega$.

[G] $\Sigma_0 \vdash \Omega$.

Proof: Since $B_n \in \Sigma_0$ for all n, we have: $\Sigma_0 \vdash [B_0 \wedge B_1 \wedge \ldots \wedge B_n] \equiv [A_0 \wedge (A_0 \Rightarrow A_1) \wedge ((A_0 \wedge A_1) \Rightarrow A_2) \wedge \ldots \wedge ((A_0 \wedge \ldots A_{n-1}) \Rightarrow A_n)] \vdash A_n$. Therefore: $\Sigma_0 \vdash A_n$ for all n. I.e. $\Sigma_0 \vdash \{A_0, A_1, \ldots\} = \Omega_0$. But, since Ω_0 is an axiom system of Ω, $\Omega_0 \vdash \Omega$. Hence $\Sigma_0 \vdash \Omega$.

[H] For any m and n such that $m \neq n$, $B_m \vee B_n$ is a tautology.

Proof: without any loss of generality, suppose that $m < n$, i.e. $m \leq (n-1)$. By (2): $[B_m \vee B_n] \equiv [((A_0 \wedge \ldots \wedge A_{m-1}) \Rightarrow A_m) \vee ((A_0 \wedge \ldots \wedge A_{n-1}) \Rightarrow A_n)]$. Hence:

$$\vdash [B_m \vee B_n] \Leftrightarrow [\neg A_0 \vee \neg A_1 \vee \ldots \vee \neg A_{m-1} \vee A_m \vee \neg A_0 \vee \neg A_1 \vee \ldots \vee \neg A_{n-1} \vee A_n].$$

The last disjunction is a tautology; for since $m \leq n - 1$, the last square bracket contains the two disjuncts A_m and $\neg A_m$. Hence result.

[I] Let $(H_0, H_1, H_2, \ldots)$ be the subsequence of $(B_0, B_1, B_2, \ldots)$ consisting of all the (distinct) B_i's which are not logical truths. Put $\Sigma = \{H_n: n \in \omega\}$. Thus Σ is obtained from Σ_0 by striking out all the logically true members of Σ_0. (Note that the set Σ need not be recursive). By **[H]**: for any $m \neq n$, $H_m \vee H_n$ is a tautology.

[J] $\Sigma \subseteq \Omega$ and $\Sigma \vdash \Omega$.

Proof: By definition, $\Sigma \subseteq \Sigma_0$. Hence, by **[F]** above, $\Sigma \subseteq \Omega$. Moreover: $\Sigma \vdash \Sigma_0$, because Σ is obtained by eliminating the *logically true* elements of Σ_0. But by [G], $\Sigma_0 \vdash \Omega$. Therefore: $\Sigma \vdash \Omega$. It remains for us to show that:

[K] Σ is logically independent.

Proof: Suppose otherwise, and let H_u be such that $(\Sigma - \{H_u\}) \vdash H_u$. By the compactness theorem, there exists a finite subset Δ of $(\Sigma - \{H_u\})$ such that $\Delta \vdash H_u$. Since $\Delta \subseteq (\Sigma - \{H_u\})$, Δ is of the form: $\Delta = \{H_{q(1)}, H_{q(2)}, \ldots, H_{q(s)}\}$, where: q(1), q(2),...,q(s) are distinct both from each other and from u. Since $\Delta \vdash H_u$, we have, by the deduction theorem: $\vdash [(H_{q(1)} \wedge \ldots \wedge H_{q(s)}) \Rightarrow H_u]$. But by [I] above, $H_{q(i)} \vee H_u$ is a tautology for each i = 1, 2,..., s. Therefore: $\vdash [H_{q(1)} \vee H_u]$, $\vdash [H_{q(2)} \vee H_u]$,..., and $\vdash [H_{q(s)} \vee H_u]$. Together with $\vdash [(H_{q(1)} \wedge \ldots \wedge H_{q(s)}) \Rightarrow H_u]$, this yields $\vdash H_u$; i.e. H_u is a logical truth, contrary to the definition of $\{H_o, H_1, \ldots\}$. Hence Σ is logically independent.

APPENDIX II

Natural Axiomatisation (A Revision of Wajsberg's Requirement)

Let *S* be a first-order system, and *L* the language of *S*. We suppose that *L* is countable and that for any truth-function p, *L* possesses a connective **p** corresponding to p. Where no confusion can arise, **p** will be identified with p. (Note that for any $n \in \omega$, there is only a finite number, namely $\exp(2,2^n)$, of truth-functions having n argument-places. Keep n fixed for the time being. After lexicographically ordering the 2^n n-tuples which constitute the arguments of these functions, the latter can, in their turn, be similarly ordered. Denote by $k_{m,,n}$ the mth truth-function, i.e. *connective*, which has n argument-places. Since m and n are natural numbers, the set $\{k_{m,n}: m < \exp(2,2^n)\}$ is denumerable, which is consistent with our assumption that *L* is countable. For $n = 0$, we have $m < \exp(2,2^0) = \exp(2,1) = 2$. We take $k_{0,0}$ and $k_{1,0}$ to be the truth-values False and True respectively. Denote $k_{0,0}$ by *f* and $k_{1,0}$ by *t*. Call a well-formed formula (wff) a propositional unit if it is either atomic or starts with a quantifier, i.e. with $(\forall$ or with $(\exists$. Thus every wff W is built from propositional units by means of the $k_{m,n}$'s; these propositional units are called the propositional components of W.

Let $A = \{A_i: i \in J\}$ be an axiom system of *S*. Denote by Ω the set of all propositional components of members of *A*. As far as assignments of truth-values are concerned, syntactically different elements of Ω, e.g. $(\forall x)P(x)$ and $(\forall x)(P(x) \wedge P(x))$, will be treated as different entities. A sentence Z whose propositional components lie in Ω is said to be *tautological* if it turns out true under all assignments of truth-values to members of Ω. In such a case, we also have, by completeness, $\vdash$ Z. X~Y will henceforth mean that the sentences X and Y are tautologically equivalent; i.e. that $X \Leftrightarrow Y$ is a tautology; which entails $\vdash (X \Leftrightarrow Y)$.

A Revision of Wajsberg's Requirement

Consider $A_j \in \mathcal{A}$. Let $C_1, C_2, \ldots, C_n$ be the sequence—from left to right—of all the distinct *occurrences* of members of Ω in the axiom A_j. Thus there exists a connective f with n arguments, such that $A_j \sim f(C_1, C_2, \ldots, C_n)$. The axiom A_j is said to be *impermissible* if there are a permutation $< i(1), i(2), \ldots, i(n) >$ of $<1, 2, \ldots, n>$, a number m such that $1 \leq m < n$, and connectives g and h with (respectively) m + 1 and n – m argument-places, such that:

(1) $f(C_1, \ldots, C_n) \sim g(C_{i(1)}, \ldots, C_{i(m)}, h(C_{i(m+1)}, \ldots, C_{i(n)}))$, where $h(C_{i(m+1)}, \ldots, C_{i(n)})$ is a theorem of *S* and hence belongs to *S* (or equivalently: $\mathcal{A} \vdash h(C_{i(m+1)}, \ldots, C_{i(n)})$, since $\mathcal{A}$ is taken to be an axiom system for *S*.) Thus: $A_j \sim f(C_1, \ldots, C_n) \sim g(C_{i(1)}, \ldots, C_{i(m)}, h(C_{i(m+1)}, \ldots, C_{i(n)}))$.

A is said to be propositionally *permissible* (or *natural*) if none of its elements is impermissible in the above sense. (Note that, by (1), the impermissibility of A_j is *a property* of A_j which is independent of the remaining elements of *A*, i.e. of any particular axiomatisation of *S*. Such a property could have been defined for any sentence whose propositional components lie in Ω).

Three Theorems

Theorem 1: Suppose A_j is impermissible in the above sense. Then: in the axiom system *A*, A_j can be replaced by the two separate axioms $h(C_{i(m+1)}, \ldots, C_{i(n)})$ and $g(C_{i(1)}, \ldots, C_{i(m)}, t)$.

Proof: Trivially, $h(C_{i(m+1)}, \ldots, C_{i(n)}) \Rightarrow [g(C_{i(1)}, \ldots, C_{i(m)}, t) \Leftrightarrow g(C_{i(1)}, \ldots, C_{i(m)}, h(C_{i(m+1)}, \ldots, C_{i(n)}))]$ is a tautology. Hence, by (1): $h(C_{i(m+1)}, \ldots, C_{i(n)}) \Rightarrow [g(C_{i(1)}, \ldots, C_{i(m)}, t) \Leftrightarrow A_j]$ is also tautological. Putting $h^* \equiv h(C_{i(m+1)}, \ldots, C_{i(n)})$ and $g^* \equiv g(C_{i(1))}, \ldots, C_{i(m)}, t)$, we have that $h^* \Rightarrow (g^* \Leftrightarrow A_j)$ is a tautology. Hence $\vdash h^* \Rightarrow (g^* \Leftrightarrow A_j)$. By hypothesis, i.e. by (1), $A \vdash h^*$; therefore $A \vdash g^* \Leftrightarrow A_j$; hence also $A \vdash g^*$, because $A_j \in A$. Thus: $A \vdash A^*$, where A^* is defined by: $A^* = (A - \{A_j\}) \cup \{h^*, g^*\}$. It remains for us to show the converse. This is however trivial since $\vdash h^* \Rightarrow (g^* \Leftrightarrow A_j)$ entails $\{h^*, g^*\} \vdash A_j$; therefore: $A^* = (A — \{A_j\}) \cup \{h^*, g^*\} \vdash (A — \{A_j\}) \cup \{A_j\} = A$.

Consequence: again, let A_j satisfy (1). By hypothesis, there exists—in our primitive vocabulary—a connective q which has m arguments and is such that: for all $X_1, X_2,\ldots, X_m$, $q(X_1, X_2,\ldots, X_m) \sim g(X_1, X_2,\ldots, X_m, t)$. Put: $q^* \equiv q(C_{i(1)},\ldots, C_{i(m)}) \sim g(C_{i(1)},\ldots, C_{i(m)}, t)$ and $A_0 = (A - \{A_j\}) \cup \{q^*, h^*\}$.

The above result is equivalent to: $A \vdash A_0$ and $A_0 \vdash A$. The only important addition is that, since all of $C_1,\ldots, C_n$ occur in A_j, then each of q^* and h^* is strictly shorter than A_j. If the set *A* of (proper) axioms of *S* is finite, we can start by considering an impermissible axiom A_j of maximum length and replace it by the corresponding set $\{q^*, h^*\}$; then proceed to the next remaining impermissible axiom of maximum length; and so on until we reach a (propositionally) natural set of axioms. We could finally eliminate, in turn, every axiom which logically follows from the remaining ones; we would thus obtain—in a non effective way—a finte set of permissible and logically independent axioms. This shows that the requirement of naturalness imposes no constraints on the content of the theory *S*.

Theorem 2: Suppose that A_j strictly contains—in the syntactical sense a subsentence K such that: K is a theorem of *S* and K contains at least one component C_u of A_j. Then A_j is impermissible.

Proof: To say that K occurs in A_j is to say that $K \sim h(C_r, C_{r+1},\ldots, C_s)$, where $1 \leq r \leq s \leq n$; and that $f(C_1,\ldots, C_n)$ is of the form $f(C_1,\ldots, C_n) \sim p(C_1,\ldots, C_{r-1}, h(C_r,\ldots, C_s), C_{s+1},\ldots, C_n)$. Let g be the connective with (r + n-s) argument places such that $g(X_1,\ldots,X_{r+n-s}) \sim p(X_1,\ldots, X_{r-1}, X_{r+n-s}, X_r,\ldots, X_{r+n-s-1})$. Hence: $A_j \sim f(C_1,\ldots, C_n) \sim p(C_1,\ldots, C_{r-1}, h(C_r,\ldots, C_s), C_{s+1},\ldots, C_n) \sim g(C_1,\ldots, C_{r-1}, C_{s+1},\ldots,C_n, h(C_r,\ldots, C_s))$. By (1), A_j is impermissible.

We shall now give examples both of (propositionally) impermissible and of natural aximatisation. For any propositional unit A, the axiom System {A} is trivially permissible; for in this case, we have n = 1, so there exists no natural number m such that $1 \leq m < n$.

Theorem 3: No conjunction is a permissible axiom; but for any three (independent) propositional units D, E and G, $\{D \Rightarrow E, E \Rightarrow G\}$ is a natural axiom system [We recall that a propositional unit is any sentence which is either atomic or else of the form $(\forall x)B(x)$ or $(\exists x)B(x)$].

Proof: [a] Let $A \equiv (A_1 \wedge A_2 \wedge \ldots \wedge A_n)$, where we assume that A is an axiom of *S*. Thus every A_j is a theorem of *S*. We can write $A \sim g(A_1 \wedge \ldots \wedge A_{n-1} \wedge h(A_n))$, where g and h are connectives such that $g(X_1,\ldots, X_n) \sim (X_1 \wedge \ldots \wedge X_n)$ and $h(X) \sim X$. Since $h(A_n)$, i.e A_n, is a theorem of *S*, it follows by (1) that A is impermissible.

[b] Consider $\{D \Rightarrow E, E \Rightarrow G\}$, where D, E, and G are propositional units. Hence, $\Omega = \{D, E, G\}$. Suppose that $\{D \Rightarrow E, E \Rightarrow G\}$ is an unnatu-

ral axiom system of *S*. It suffices to derive a contradiction. By definition: D ⇒ E or E ⇒ G must be impermissible. Assume that D ⇒ E is impermissible. By (1) above, we must have some m such that $1 \leq m < 2$. Hence m = 1. So there are only two possible subcases. The first is where (D ⇒ E) ~ g(D,h(E)) and h(E) is a theorem of *S*; i.e. h(E) is a logical consequence of {D ⇒ E, E ⇒ D}. Give both D and E the truth-value *t* (i.e. True). Thus both elements of {D ⇒ E, E ⇒ D} are true; which entails that h(E) is true for E true. I.e. h(*t*) ~ *t*. Now give both D and E the value *f* (i.e. False). Thus both elements of {D ⇒ E, E ⇒ D} are true, so that h(E) must also be true; hence h(E) is true for E false; i.e. h(*f*) ~ *t*. Hence h(E) is true for all truth-values of E; i.e h(E) ~ *t*. Therefore (D ⇒ E) ~ g(D,h(E)) ~ g(D,*t*). We can now easily derive a contradiction from the fact that the truth-value of (D⇒E) depends only on that of D: taking both D and E to be true, obtain (*t* ⇒ *t*) ~ g(*t, t*); i.e. *t* ~ g(*t, t*). Taking D to be true and E false, we have (*t* ⇒ *f*)~ g(*t, t*); i.e. *f* ~ g(*t, t*); therefore *t* ~ *f*; whence a contradiction.

In the second subcase, we have (D ⇒ E) ~ g(E,h(D)), where h(D) logically follows from {D ⇒ E, E ⇒ D}. Taking both D and E to be true, then both D and E to be false, we conclude that h(D) is always true; i.e. h(D) ~ *t*. Hence: (D ⇒ E) ~ g(E,h(D)) ~ g(E,*t*); so that the truth-value of (D ⇒ E) now depends only on that of E. Taking the truth-values of both D and E to be false, we conclude that *t* ~ *g(f, t)*. Taking D to be true and E false, we obtain *f* ~ g(*f, t)*. Hence: *t* ~ *f*, which yields a contradiction. The case where (E ⇒ D) is impermissible is similarly treated. Hence result.

(Note however that in the proof above, we have implicitly assumed that D, E and G are 'independent' of each other; more precisely: that an interpretation can be found under which any subset of {D,E, G}, but not its complement, is verified. This would not be the case if e.g. D ≡ (∀x)F(x) and E ≡ F(a).)

The Problem of Quantification

As presented in Watkins's 1984, Wajsberg's requirement had a clause extending it to cases where an axiom A_j contains a component which becomes a theorem only after we bind its variables by the same quantifiers binding them in A_j. An example given by Watkins (with acknowledgements to Glymour) of an unnatural axiomatisation involving quantifiers is the following. Let *S* = <(∀x)[(D(x) ⇒ E(x)) ∧ (E(x) ⇒ G(x))]> = consequence class of (∀x)[(D(x) ⇒ E(x)) ∧ (E(x) ⇒ G(x))].

An intuitively natural axiomatisation of S *would be* $\{(\forall x)(D(x) \Rightarrow E(x)), (\forall x)(E(x) \Rightarrow G(x))\}$; while an intuitively unnatural one would be: $\{A_1, A_2\}$, where A_1 and A_2 are defined by: $A_1 \equiv (\forall x)[(D(x) \Rightarrow G(x)) \Rightarrow ((D(x) \Rightarrow E(x)) \wedge (E(x) \Rightarrow G(x)))]$, $A_2 \equiv (\forall x)(D(x) \Rightarrow G(x))$.

We do not as yet have anything with which to exclude such unnatural axiomatisations; for a sentence starting with $(\forall$ or with $(\exists$ counts as a propositional unit.

The following is a tentative way of extending the above definition of impermissibility to quantified sentences. Without any loss of generality, suppose that our primitive vocabulary contains no function symbols. Introduce a set of distinct names $\{a_i: i \in \Delta\}$ for all the elements of our domain of discourse whose cardinality must therefore be equal to that of Δ. Take every assumption of the (purely universal) form $(\forall x_1 \ldots x_n)B(x_1, \ldots, x_n)$ to represent the—possibly infinite—set of axioms: $\{B(a_{j(1)}, \ldots, a_{j(n)}): <j(1), \ldots, j(n)> \in \Delta^n\}$. Having carried out this operation for every axiom, examine whether or not the new set of unquantified axioms is permissible in the sense of (1) above. In the case of the Glymour example, we obtain: $\{[(D(a_i) \Rightarrow G(a_i)) \Rightarrow ((D(a_i)) \Rightarrow E(a_i)) \wedge (E(a_i) \Rightarrow G(a_i)))], D(a_i) \Rightarrow G(a_i): i \in \Delta\}$.

This set has now become impermissible, for $D(a_i) \Rightarrow E(a_i)$, $E(a_i) \Rightarrow G(a_i)$ and $D(a_i) \Rightarrow G(a_i)$ are all theorems of *S*. We now use Theorem 1 and its consequence in order to work back to a 'natural' axiomatisation. Substituting the tautology *t* for $D(a_i) \Rightarrow E(a_i)$ and then adding $D(a_i) \Rightarrow E(a_i)$ to our axioms, we have: $\{D(a_i) \Rightarrow E(a_i), [(D(a_i) \Rightarrow G(a_i)) \Rightarrow (t \wedge ((E(a_i) \Rightarrow G(a_i)))], D(a_i) \Rightarrow G(a_i): i \in \Delta\}$. We similarly have to replace $E(a_i) \Rightarrow G(a_i)$ by *t* and then adjoin $E(a_i) \Rightarrow G(a_i)$ to our axioms, which yields: $\{[D(a_i) \Rightarrow E(a_i)], [E(a_i) \Rightarrow G(a_i)], [(D(a_i) \Rightarrow G(a_i)) \Rightarrow (t \wedge t)], [D(a_i) \Rightarrow G(a_i)]: i \in \Delta\}$.

Of these axioms, the third is now a tautology, and the fourth a logical consequence of the first two; they must therefore both be dropped in accordance with the requirement of logical independence, thus leaving: $\{D(a_i) \Rightarrow E(a_i), E(a_i) \Rightarrow G(a_i): i \in \Delta\}$. Re-introducing the quantifiers, we retrieve the initial set of intuitively 'natural axioms: $\{(\forall x)(D(x) \Rightarrow E(x)), (\forall x)(E(x) \Rightarrow G(x))\}$.

There remains the more difficult case of mixed quantifiers. Here, it is useful to introduce Skolem functions by first expressing each axiom in the equivalent second-order form: $(\exists f_1 \ldots f_m)(\forall x_1 \ldots x_n)B(f_1, \ldots, f_m; x_1, \ldots, x_n)$ where, for each index u, f_u is a function variable with, say, q_u argument-places. Let $g_1, \ldots, g_m$ be distinct and new, i.e. hitherto unused function constants where, for each u, g_u has q_u argument-places. Replace the above axiom by the (trivially) stronger first-order sentence:

$(\forall x_1 \ldots x_n)B(g_1, \ldots, g_m; x_1, \ldots, x_n)$. We are however interested exclusively in the consequences of such sentences in which $g_1, \ldots, g_m$ do not occur; and these already follow from the old axioms. Qua physical theory, the new sentence will consequently be equivalent to the old one.

Proceeding as above, we substitute for the last sentence the generally infinite set of axioms: $\{B(g_1, \ldots, g_m; a_{j(1)}, \ldots, a_{j(n)}): \langle j(1), \ldots, j(n) \rangle \in \Delta^n\}$. As explained in the case of Glymour's example, this may enable us to apply the (propositional) impermissibility criterion and—where possible—to construct an allowable axiom system through re-introducing the quantifiers.

Let us reconsider the theory *S* unnaturally axiomatised by {H} where: $H \equiv (\forall x)(\exists y)(\forall v)(\exists z)(\forall u)[R(x, y) \wedge (R(z, u) \Rightarrow P(v))]$ (see above, p. 223). Introducing the Skolem functions f and g, we replace H by the assumption: $(\forall x)(\forall v)(\forall u)[R(x, f(x)) \wedge (R(g(x, v), u) \Rightarrow P(v))]$. We now go over to the infinite set of quantifier-free postulates: $\{R(a_1, f(a_1)) \wedge (R(g(a_1, a_2), a_3) \Rightarrow P(a_2)): \langle a_1, a_2, a_3 \rangle \in \Delta^3\}$ (where Δ now denotes a set of names for all the elements of our domain of discourse). This axiomatisation is inadmissible because it contains conjunctions of various formulas. Applying the method described in Theorem 1, we substitute *t* for $R(a_1, f(a_1))$, which we then add as a separate axiom, thus obtaining: $\{R(a_1, f(a_1)), t \wedge (R(g(a_1, a_2), a_3) \Rightarrow P(a_2)): \langle a_1, a_2, a_3 \rangle \in \Delta^3\}$.

Once again, the second axiom is a conjunction, so that, proceeding as above, we have: $\{R(a_1, f(a_1)), t \wedge t, R(g(a_1, a_2), a_3) \Rightarrow P(a_2): \langle a_1, a_2, a_3 \rangle \in \Delta^3\}$.

The requirement of logical independence leads us to the following set: $\{R(a_1, f(a_1)): a_1 \in \Delta\} \cup \{R(g(a_1, a_2), a_3) \Rightarrow P(a_2): \langle a_1, a_2, a_3 \rangle \in \Delta^3\}$. Re-introducing the quantifiers, we obtain: $\{(\forall x)R(x, f(x))\} \cup \{(\forall x)(\forall v)(\forall u)[R(g(x, v), u) \Rightarrow P(v)]\}$. Since we are interested only in the elements of *S* containing neither f nor g, we can replace the last axiom-set by: $\{(\forall x)(\exists y)R(x, y)\} \cup \{(\forall x)(\forall v)(\exists z)(\forall u)[R(z, u) \Rightarrow P(v)]\}$. I.e.: $\{(\forall x)(\exists y)R(x, y), (\forall x)(\exists y)R(x, y) \Rightarrow (\forall v)P(v)\}$.

This axiom set is impermissible because $(\forall x)(\exists y)R(x, y)$ is both a theorem of *S* and a component of $(\forall x)(\exists y)R(x, y) \Rightarrow (\forall v)P(v)$. In this last sentence, we therefore have to substitute *t* for $(\forall x)(\exists y)R(x, y)$ which must then be added as an axiom. This yields: $\{(\forall x)(\exists y)R(x, y), (\forall x)(\exists y)R(x, y), t \Rightarrow (\forall v)P(v)\}$.

For the same reason and through applying a similar method, we obtain: $\{(\forall x)(\exists y)R(x, y), (\forall x)((y)R(x, y), (\forall v)P(v), t \Rightarrow t\}$. The condition of logical independence entitles us to delete the first and the fourth axioms, thus leaving the intuitively natural axiomatisation: $\{(\forall x)(\exists y)R(x, y), (\forall v)P(v)\}$ (see p. 000).

APPENDIX III

The Empirical Unity of Scientific Systems and Watkins's Organic Fertility Requirement (OFR)

As already mentioned, the problem of the unity of scientific hypotheses has been one of the scientists' and epistemologists' main concerns. Although the simplicity of high-level theories might ultimately turn out to be an unaffordable luxury, the unity requirement remains—at any rate according to Poincaré—a transcendental presupposition of science as such. For their part, both Einstein and Popper spoke of the organic compactness of fundamental theories, while wondering whether this notion is susceptible of ever being rigorously defined. Einstein alluded to the paucity of the primitive elements occurring in a theory as to a measure of the latter's degree of unity. But unity has intuitively to do, not with a theory's basic vocabulary, but with the connections between the various components of an axiom system; the latter must be such that none of its subclasses is, *in a sense yet to be defined*, isolable from the rest.

In his 1984, John Watkins put forward the view that a hypothesis is unified if it satisfies an Organic Fertility Requirement (OFR), that is: if *all* of its postulates collaborate towards yielding empirical results not entailed by any subclass of its axioms. More precisely, Watkins starts by defining the empirical content Ct(S) of a theory S as the set of all propositions of the form e $\Rightarrow$ p, where: $S \vdash$ (e $\Rightarrow$ p), p is a prediction, and e a description of the relevant boundary conditions. Let E be the class of all the axioms of S, where E is taken to be finite. Thus $S = <\Lambda E>$, where, for any (finite) set of formulas G, ΛG denotes the conjunction of all members of G. Consider any non-trivial partition (E_1, E_2) of E; i.e.: $E_1 \neq \varnothing \neq E_2$, $E = E_1 \cup E_2$, $E_1 \cap E_2 = \varnothing$.

Watkins's OFR was initially taken to be fufilled if

(*) $\mathrm{Ct}(S) \neq \mathrm{Ct}(\Lambda E_1) \cup \mathrm{Ct}(\Lambda E_2)$, i.e. if $\mathrm{Ct}(\Lambda E) \neq \mathrm{Ct}(\Lambda E_1) \cup \mathrm{Ct}(\Lambda E_2)$.

Since $E_1 \subset E$, then $\vdash (\Lambda E) \Rightarrow (\Lambda E_1)$. From an intuitive standpoint, we must therefore require $\mathrm{Ct}(S) = \mathrm{Ct}(\Lambda E) \supseteq \mathrm{Ct}(\Lambda E_1)$. Similarly: $\mathrm{Ct}(S) \supseteq \mathrm{Ct}(\Lambda E_2)$. Therefore: $\mathrm{Ct}(S) \supseteq \mathrm{Ct}(\Lambda E_1) \cup \mathrm{Ct}(\Lambda E_2)$. Hence (*) will be equivalent to: (**) $\mathrm{Ct}(S) \supset \mathrm{Ct}(\Lambda E_1) \cup \mathrm{Ct}(\Lambda E_2)$.

It seems to me that Watkins's criterion adequately characterises what is intuitively meant by the *empirical* unity of a scientific theory *S*. (*), however, depends on a pregiven definition of Ct(*S*) and on the mode *E* of axiomatising *S*. Let us start with the second type of dependence. Take K to be any metaphysical theory which is empirically non-verifiable and non-refutable. Put $B \equiv \Lambda E$. Thus one possible axiom system of *S* is: $\{B \vee K, B \vee \neg K\}$, where both $B \vee K$ and $B \vee \neg K$ are experimentally unfalsifiable because K is neither refutable nor verifiable. Hence: $Ct(B \vee K) = \varnothing = Ct(B \vee \neg K)$. Since B is scientific, then by definition: $Ct(B) \neq \varnothing$. Therefore: $Ct(S) = Ct(B) \supset \varnothing = Ct(B \vee K) \cup Ct(B \vee \neg K)$. By (*), it would thus follow that any arbitrary scientific hypothesis *S* is unified, which trivialises Watkins's OFR. Note however that the axiomatisation $\{B \vee K, B \vee \neg K\}$ is impermissible; for B is both a theorem of $S = \langle B \rangle$ and a component of both axioms $B \vee K$ and $B \vee \neg K$ (see Theorem 2, Appendix 2). We shall therefore restrict (*) to the class of all natural axiomatisations of *S*.

Let us now turn to Watkins's definition of the empirical content Ct(*S*) of *S*, a definition which poses serious problems; for consider a non-unified theory *S* consisting of the axiom-set $\{B_1, B_2\}$, where B_1 and B_2 possess disjoint domains. We have: $S = \langle B_1 \wedge B_2 \rangle$. Let $e_1 \Rightarrow p_1$ and $e_2 \Rightarrow p_2$ be predictions of B_1 and B_2 respectively. Thus, by definition: $B_1 \vdash (e_1 \Rightarrow p_1)$ and $B_2 \vdash (e_2 \Rightarrow p_2)$. Furthermore, according to Watkins: $(e_1 \Rightarrow p_1) \in Ct(B_1)$ and $(e_2 \Rightarrow p_2) \in Ct(B_2)$. Since $B_1 \wedge B_2 \vdash (e_1 \wedge e_2) \Rightarrow (p_1 \wedge p_2)$, then, invoking Watkins's criterion, we must have $[(e_1 \wedge e_2) \Rightarrow (p_1 \wedge p_2)] \in Ct(B_1 \wedge B_2) = Ct(S)$. But since B_1 and B_2 have nothing to do with each other, we normally have both Not$[B_1 \vdash (e_1 \wedge e_2) \Rightarrow (p_1 \wedge p_2)]$ and Not$[B_2 \vdash (e_1 \wedge e_2) \Rightarrow (p_1 \wedge p_2)]$; from which $[(e_1 \wedge e_2) \Rightarrow (p_1 \wedge p_2)] \notin Ct(B_1)$ and $[(e_1 \wedge e_2) \Rightarrow (p_1 \wedge p_2)] \notin Ct(B_2)$ follow. Therefore: $[(e_1 \wedge e_2) \Rightarrow (p_1 \wedge p_2)] \notin [Ct(B_1) \cup Ct(B_2)]$ and so $Ct(B_1 \wedge B_2) \neq [Ct(B_1) \cup Ct(B_2)]$; i.e. $Ct(S) \neq [Ct(B_1) \cup Ct(B_2)]$. Should the axiomatisation $\{B_1, B_2\}$ be natural, then we must conclude that *S* is unified; which is highly counter-intuitive since B_1 and B_2 have, by hypothesis, disjoint domains of discourse. Apart from naturalness, we thus have a good reason for requiring Ct(*S*) to be logically independent; for if $\{e_1 \Rightarrow p_1, e_2 \Rightarrow p_2\} \subseteq Ct(S)$, then since $\{e_1 \Rightarrow p_1, e_2 \Rightarrow p_2\} \vdash (e_1 \wedge e_2) \Rightarrow (p_1 \wedge p_2)$, logical independence would entail that $[(e_1 \wedge e_2) \Rightarrow (p_1 \wedge p_2)] \notin Ct(S)$. We shall now put forward a number of adequacy requirements for Ct(*S*), without however being able to define Ct(*S*) explicitly.

1. We have just seen why Ct(*S*) ought to be logically independent.

2. As already noted, we also expect the empirical content of *S* to grow, or at least not to decline, with the logical strength of *S*. Hence we require $S \vdash U$ to entail $Ct(S) \supseteq Ct(U)$.

3. Let Et(S) be the set of all (fully) empirically decidable members of S. One requirement must be that: Ct(S) $\subseteq$ Et(S), for we naturally want all elements of Ct(S) to be empirically decidable. (We must in fact have Ct(S) $\subset$ Et(S), since Et(S) is clearly logically dependent: for if q $\in$ Et(S) and r $\in$ Et(S), then $q \wedge r$ is also fully decidable and follows from S; i.e. $q \wedge r \in$ Et(S).)

4. We do not however want Ct(S) to be an arbitrary logically independent subset of Et(S); for we also require Ct(S) to be empirically complete relatively to S; i.e. Ct(S) should have the same empirically decidable consequences as S. This is one reason for entering the condition: Ct(S) $\vdash$ Et(S).

5. We finally want the elements of Ct(S) to be as irreducible, that.is: as nearly atomic as possible. Hence no element of Ct(S) should 'contain' any other element of Et(S), i.e. another empirically decidable element of S. Every member b of Ct(S) must therefore satisfy a Wajsberg-type condition. In other words: b should not be of the form: $b \sim g(d_1,\ldots,d_m, h(d_{m+1},\ldots, d_n))$, where: g and h are connectives; $d_1,\ldots, d_n$ are all empirically decidable, and $h(d_{m+1}, d_{m+2},\ldots, d_n) \in$ Et(S), i.e. $S \vdash h(d_{m+1}, d_{m+2},\ldots, d_n)$; for if this were the case, then, within Ct(S), {b} could be replaced by $\{q(d_1,\ldots,d_m), h(d_{m+1},\ldots, d_n)\}$, where the connective q is such that: $q(X_1,\ldots, X_m) \sim g(X_1,\ldots, X_m, t)$ (see above, Theorem 1 and Consequence).

Note: the above do not constitute a definition of Ct(S), but a list of adequacy conditions to be imposed on the structure of Ct(S). We therefore have no guarantee that any set Ct(S) satisfying all these conditions exists.

APPENDIX IV

Ramseyfication and Structural Realism (Co-written with John Worrall)

Several authors have recently highlighted what they take to be a major objection to 'structural realism'—an account of the status of scientific theories which, as indicated in the text, was advocated by Poincaré and also by Russell. This objection is *so* major that it raises questions about the very 'possibility' of structural realism. The objection goes, essentially, as follows: (i) Structural Realism is in effect committed to the view that the cognitive content of any scientific theory T is fully captured by its Ramsey sentence T^* —where T^*, as is well known, is formed from T by first substituting variables for all the theoretical predicates involved in T and then existentially geneneralising over all those predicate-variables. (We could remain within first-order logic by regarding all relations as classes of n-tuples of individuals; but then we would have to look upon the notions of a class ($k(x) \equiv x$ is a class) and of an object belonging to a class (($y \in x$) $\equiv$ y belongs to the class x) as logical constants; so that k and $\in$ ought not to be quantified over when we form T^*. It is worth noting that in the—very simple—examples adduced below, all existential quantifications over predicate variables will anyway yield formulas logically equivalent to first-order sentences.)

(ii) But, as was allegedly first shown by the mathematician M.H.A. Newman—responding to Russell's version of structural realism—a theory's Ramsey-sentence is satisfiable in a 'trivial' way, imposing no more, supposedly, than a cardinality constraint (see Newman 1928).

(iii) However, scientific theories clearly do much more than impose constraints on the size of the universe.

(iv) So stuctural realism is committed to an entirely untenable account of the cognitive content of scientific theories—that that content is captured by the theory's Ramsey-sentence—and is itself untenable.

Newman's argument was reintroduced into the philosophy of science by Demopoulos and Friedman—and via their article, has attracted a good deal of attention in the past few years (Demopoulos and Friedman 1985). We shall however show that despite being seemingly cogent, Newman's objections owe their *superficial* cogency to an oversight, or rather to an easily corrigible mistake on Russell's part. After rectifying Russell's error, it will be seen that Newman's argument in no way threatens structural realism—though consideration of this argument will give us the opportunity to clarify further certain important aspects of the position (including that of whether or not structural realism is really committed to the view that the cognitive content of any theory is captured by that theory's Ramsey-sentence).

[A] Ramseyfication and 'Trivialisation'

Demopoulos and Friedman quote Russell's claim, made in his1927, that:

> Whatever we infer from perceptions, it is only structure that we can validly infer; and structure is what can be expressed by mathematical logic . . . (Russsell 1927, p. 254)

> The only legitimate attitude about the physical world seems to be one of complete agnosticism as regards all but its mathematical properties. (Russell 1927, p. 270)

According to Demopoulos and Friedman, Newman, who explicitly referred to Russell, had shown that any view which identifies the cognitive content of a scientific theory with that of its Ramsey sentence faces "insurmountable difficulties." The crucial paragraph of their paper, which in turn quotes Newman, runs as follows:

> The difficulty is with the claim that *only* structure is known. On this view "the world consists of objects, forming an aggregate whose structure with respect to a certain relation R is known, say [it has] structure W; but of R nothing is known . . . but its existence; . . . all we can say is *There is* a relation R such that the structure of the external world with reference to R is W" (Newman 1928, p. 144). But "*any* collection of things can be organised so as to have the the structure W, provided there are the right number of them" (p. 144; italics added). Thus, on this view, only cardinality questions are open to discovery! Every other claim about the world that can be known at all can be known a priori as a logical consequence of the existence of α-many objects. For any given set A of cardinality α can, with a minimum amount of set theory or second-order logic, establish the existence of a relation having the structure W, provided that W is compatible with the cardinality constraint that $|A| = \alpha$. (The relevant theorem from set theory or second-order logic is the proposition

> that every set A determines a full structure, that is one which contains every relation [in extension] of every arity on A; such a structure forms the basis for a [standard] model for the language of second- [or higher-] order logic). (Demopoulos and Friedman 1985, pp. 627–28).

It was admittedly unfortunate that in his 1927 Russell spoke of a *purely* structural description of reality being *inferred* from perceptual results. The fault lies not so much with the inductive overtones of the verb 'infer', invalid though any such inference might be; as with the implicit assumption that once the inference is carried out, an exclusively structural account is obtained in which no observational terms occur. After being used during the inferential process, all observations are supposedly shed, thus yielding a structural proposition $H(T_1,\ldots,T_m)$ in which all the predicates $T_1,\ldots,T_m$ are theoretical [$H(T_1,\ldots,T_m)$ will henceforth be assumed to be a first-order sentence.] Ramseyfying, we obtain $(\exists t_1,t_2, \ldots,t_m)H(t_1,\ldots,t_m)$; where $(\exists t_1,t_2, \ldots,t_m)H(t_1,\ldots,t_m)$ expresses a consistency condition for $H(T_1,\ldots,T_m)$; i.e. it means that $H(T_1,\ldots,T_m)$ possesses at least one model whose domain can moreover be taken to be countable; provided of course that $H(T_1,\ldots,T_m)$ is consistent (see below Section 2, case [b].) Consistently with his overall philosophical position, Russell could however have adopted an alternative, more precisely: a hypothetico-deductivist approach. He would then have put forward a mixed empirico-structural proposition $G(Q_1,\ldots,Q_n,T_1,\ldots,T_m)$ where $Q_1,\ldots,Q_n$ are observational terms. Being empirically falsifiable, the corresponding Ramsey-sentence, namely $(\exists t_1,t_2,\ldots,t_m)G(Q_1,\ldots,Q_n,t_1,\ldots,t_m)$, would then have been far from trivially satisfiable (see below, Section 2). Note however that this defence of Russell's structuralist thesis is predicated on a division between 'theoretical' and 'observational' terms; where the latter are taken to be understood in advance of any physical theory, i.e. directly and without any involvement of description. This means that those 'antecedently understood' predicates retain their meaning (that is, in extension) from interpretation to interpretation. If structural realism did indeed imply that *all* that can be 'known' is stucture, then it would face insurmountable difficulties; but in fact the view is that while *theoretical* entities are known only via description, observational entities are known both by description and by ostension, or whatever. It should moreover be noted that in the global description $G(Q_1,\ldots,Q_n,T_1,\ldots,T_m)$, the theoretical T's are indissolubly entangled with the observational Q's.

Nowadays, it is generally held that there is no absolute or sharp distinction between theoretical and observational predicates. Russell (along with Poincaré, as well as Grover Maxwell, Ayer, and the author of the present book) would—to the contrary—happily identify the observational predicates with those decidable 'phenomenologically' on the basis of perception—for Russell we of course have knowledge by *acquaintance* as well as by description. (Indeed, his structural realism simply consists of thinking through the claim that while we do have knowledge by acquaintance of percepts, we have knowledge of theoretical entities only by description; see above, 1.7; also Maxwell 1971).

We need not however commit ourselves to any particular account of the 'observational basis' of science here. Some may wish to follow David Lewis in taking the distinction to be simply one between 'old' predicates (taken—within the current context—*somehow or other* to be antecendently understood), and 'new' predicates. Others might like to take the antecedently understood, observationally decidable predicates as those involved in Poincaré's 'crude facts'—such as the fact that 'two points (for example the end of a meter needle and a certain point on the dial of the meter) roughly coincide'. But however exactly the distinction is understood, no serious version of structural realism can get going without some such distinction. If *all* the predicates of a scientific theory are taken to be interpreted only within the context of the claims made by the theory, if, that is, none is taken to be firmly anchored in experience independently of our attempted descriptions of the universe, then the constraints imposed by the Ramsey-sentence (or indeed by the original theory itself) would be hopelessly weak. For the theory would then be indistinguishable from a formal mathematical system, where the latter is—quite properly—subject to a mere consistency condition. The criticism of those who object to structural realism would then reduce to the ungainsayable truism that a consistent formal theory constitutes an 'implicit definition' of its primitive concepts; i.e. it merely determines a class of models (see Lewis 1970).

Let us immediately grant that Demopoulos and Friedman seem to be aware of this possible counter-objection when they maintain that:

> More precisely, if our theory is consistent and if all its purely observational consequences are true, then the truth of the Ramsey-sentence *follows* as a theorem of set-theory or second-order logic, provided our initial domain has the right cardinality—if it doesn't, then the consistency condition of our theory again implies the existence of a domain that does. (Demopoulos and Friedman 1985, p. 635)

Thus, according to Demopoulos and Friedman, only what the Ramsey-sentence asserts *over and above its observational content* is reducible to logic or to mathematics; more precisely: to a consistency condition or to a cardinality constraint. This 'over and above' however seems to us to be not only badly defined but also indefinable in any non-trivial way. For on the one hand, the Ramsey-sentence does not in general logically follow from its empirical basis—from the set of true, empirically decidable and hence singular atomic sentences. On the other hand, if the so-called empirical generalisations—which are generally undecidable—were to be included in the observational content of a theory, then the Ramsey-sentence might well be one of them; in which case Demopoulos's and Friedman's thesis would collapse into the trivial claim that the Ramsey-sentence follows from itself. (Needless to say, we follow Popper in holding every basic statement to be not only a singular, but also an empirically decidable proposition.) For any hypothesis H, H^* will henceforth denote the Ramsey-sentence of H. Consider the following—admittedly simplistic—example. Put $A \equiv [(\forall x)(F(x) \Rightarrow T(x)) \wedge (\forall y)(T(y) \Rightarrow K(y))]$, where T is a theoretical, while both F and K are observational concepts. Thus $A^* \equiv (\exists t)[(\forall x)(F(x) \Rightarrow t(x)) \wedge (\forall y)(t(y) \Rightarrow K(y))]$. It is obvious that $\vdash A^* \Leftrightarrow (\forall x)(F(x) \Rightarrow K(x))$. Let $\{a_j\!: j \in \Delta\}$ be a set of names for all the observable entities in the world. Δ must therefore be infinite. Assume that $F(a_j) \wedge K(a_j)$ has been ascertained as true for all $j \in \Delta$. If we are to believe Demopoulos and Friedman, what the Ramsey sentence asserts beyond $\{F(a_j) \wedge K(a_j)\!: j \in \Delta\}$ is merely tautological. This claim can be easily refuted as follows. First note that the *logical* difference between two statements Y and X—i.e. the weakest formula which, taken in conjunction with X, entails Y—is $(\neg X \vee Y)$ or $(X \Rightarrow Y)$; but here, we are not dealing with the difference between A* and a single statement, but with the difference between A* and the infinite set of observation reports $\{F(a_j) \wedge K(a_j)\!: j \in \Delta\}$. For the sake of argument, let us however suppose that the infinite conjunction $\Lambda\{F(a_j) \wedge K(a_j)\!: j \in \Delta\}$ is syntactically well-formed. Demopoulos and Friedman's thesis is thus tantamount to the claim that

$\Lambda\{F(a_j) \wedge K(a_j)\!: j \in \Delta\} \Rightarrow A^*$ is a logical truth; i.e. that

$\{F(a_j) \wedge K(a_j)\!: j \in \Delta\} \vdash (\forall x)(F(x) \Rightarrow K(x))$. That this inference is invalid can most easily be established by invoking the compactness theorem, which yields: $\{F(a_{q(1)}) \wedge K(a_{q(1)}), \ldots, F(a_{q(n)}) \wedge K(a_{q(n)})\} \vdash (\forall x)(F(x) \Rightarrow K(x))$ for some n; hence

$\vdash [(F(a_{q(1)}) \wedge K(a_{q(1)}) \wedge \ldots \wedge F(a_{q(n)}) \wedge K(a_{q(n)})) \Rightarrow (\forall x)(F(x) \Rightarrow K(x))]$.

This last formula can be falsified in any domain possessing more than n elements. The Ramsey sentence $(\forall x)(F(x) \Rightarrow K(x))$ therefore goes well beyond the set $\{F(a_j) \wedge K(a_j)\!: j \in \Delta\}$ of true atomic observation reports. (This may have to do with the fact that we have no way, in the object-language, of saying that x is to range only over the denotations of the a_j's. However the above argument does not hang on the compactness theorem; for any domain having more than $|\Delta|$ elements yields a model of

$\{F(a_j) \wedge K(a_j)\colon j \in \Delta\}$ which falsifies $(\forall x)(F(x) \Rightarrow K(x))$.)

We have already mentioned that some philosophers might choose to consider the 'empirical generalisation' $(\forall x)(F(x) \Rightarrow K(x))$, the Ramsey sentence A^* itself, as belonging to the observational basis of the hypothesis *A*. This would however go against the canons of even the most liberal version of empiricism; for $(\forall x)(F(x) \Rightarrow K(x))$ fails to be fully empirically decidable. This is precisely why Schlick and other members of the Vienna Circle decided to regard all synthetic universal statements as expressing inference rules rather than genuine propositions (see Kraft 1968, pp. 121–137). It should incidentally be noted that Popper had the great merit of realising that the quantifiers occurring in a proposition like $(\forall x)(F(x) \Rightarrow K(x))$ play as important a role in determining the methodological status of the proposition as does the nature of the concepts contained in it.

[B] What Difference, if Any, Is There between a Theory X and Its Ramsey-Sentence X*?

Let us sum up what was established above: once a distinction *is* made between the fixed observational predicates and the theoretical ones, both Newman's and the Demopoulos-Friedman objections fail.

We shall now go back to the general case of a hypothesis $G(\underline{Q}, \underline{T})$ and of its Ramsey sentence $(\exists \underline{t})G(\underline{Q}, \underline{t})$, where we have written $\underline{Q}$, $\underline{T}$ and $\underline{t}$ for the tuples $(Q_1, \ldots, Q_n)$, $(T_1, \ldots, T_m)$ and $(t_1, \ldots, t_m)$ respectively. Once again, note that the observational predicates $\underline{Q}$ are *not* quantified over, for they are regarded as having referents fixed independently of G. It will be easy to prove that the Ramsey sentence $(\exists \underline{t})G(\underline{Q}, \underline{t})$ does not merely constrain the cardinality of its models—in fact, it sometimes does less than this but, generally, it does much more. Let us compare the logical strengths of the two formulas $(\exists \underline{t})G(\underline{Q}, \underline{t})$ and $G(\underline{Q}, \underline{T})$. First note that—unsurprisingly:

(1) $\vdash [G(\underline{Q}, \underline{T}) \Rightarrow (\exists \underline{t})\, G(\underline{Q}, \underline{t})]$,

that is: the original theory always entails its Ramsey-sentence. Consider next the logical difference:

(2) $[(\exists \underline{t})G(\underline{Q}, \underline{t}) \Rightarrow G(\underline{Q}, \underline{T})]$ between $G(\underline{Q}, \underline{T})$ and $(\exists \underline{t})G(\underline{Q}, \underline{t})$.

Carnap looked upon (2) as an analytic meaning-postulate (Carnap 1966, Chapter 28). To my mind, this appellation is both wrong-headed and highly misleading; for in a previous section of the same book, Carnap had defined an analytic formula as one which proves to be logically true once its defined concepts are replaced by their definiens in terms of some basic vocabulary. In (2), both $\underline{Q}$ and $\underline{T}$ are however taken to be primitive concepts. If Carnap is right, then (2) ought to be tautologous, i.e. true under all interpretations. But this is far from being generally the case, as is shown by the example of the theory *A* and of its Ramsey-sentence *A** (see Section 1. above): consider any domain in which F, K and T denote the classes F_0, K_0 and T_0 respectively, where $F_0 \subseteq K_0$ but Not($F_0 \subseteq T_0$); this yields a model of *A** which falsifies *A*. Thus a theory is essentially stronger than its Ramsey-sentence; and it seems to me that Carnap's mistake stems from the positivist prejudice that all meanings must ultimately be anchored in perceptual experience; so that the role of (2) is alleged to consist merely in conferrring meanings onto the theoretical predicates $\underline{T}$. It can in fact be easily proved that (2) is not analytic but untestable. Let us call any closed formula which contains no theoretical terms an O-sentence; and let us define a basic statement as any singular O-sentence which is fully empirically decidable. We shall show not only that (2) is empirically irrefutable—'metaphysical' in Popper's sense—but also that it entails no synthetic O-sentence. For suppose that $\vdash [((\exists \underline{t})G(\underline{Q}, \underline{t}) \Rightarrow G(\underline{Q}, \underline{T})) \Rightarrow Z]$, where Z does not contains any $\underline{T}$. Quantifying over $\underline{T}$, we obtain: $\vdash [(\exists \underline{t})\ ((\exists \underline{t})G(\underline{Q}, \underline{t}) \Rightarrow G(\underline{Q}, \underline{t})) \Rightarrow Z]$, that is $\vdash [((\exists \underline{t})G(\underline{Q}, \underline{t}) \Rightarrow (\exists \underline{t})\ G(\underline{Q}, \underline{t})) \Rightarrow Z]$; hence $\vdash Z$; i.e. Z is a logical truth. Summing up, we can therefore say that despite its metaphysical character, (2) is not tautological.

Let us now show that the theory $G(\underline{Q}, \underline{T})$ and its Ramsey-sentence $(\exists \underline{t})G(\underline{Q}, \underline{t})$ are not only observationally identical but that they furthermore entail the same O-sentences. By (1), every consequence of $(\exists \underline{t})G(\underline{Q}, \underline{t})$ is also trivially implied by $G(\underline{Q}, \underline{T})$. Conversely, let $H(\underline{Q}, \underline{T})$ be any logical consequence of $G(\underline{Q}, \underline{T})$. That is:

(3) $\vdash G(\underline{Q}, \underline{T}) \Rightarrow H(\underline{Q}, \underline{T})$.

Quantifying over $\underline{T}$, we infer:

(4) $\vdash (\forall \underline{t})[G(\underline{Q}, \underline{t}) \Rightarrow H(\underline{Q}, \underline{t})]$.

An immediate consequence of (4) is:

(5) $\vdash [(\exists \underline{t})G(\underline{Q}, \underline{t}) \Rightarrow (\exists \underline{t})H(\underline{Q}, \underline{t})]$.

Note first that (5) is in general much weaker than (4); secondly that, in (3), we could have generalised over *all* the predicates T_j *and* Q_i. We refrained from doing so precisely because the Q_i's, being observational, have fixed meanings; so that quantifying over them—as Newman seems to have done—would entail a loss of valuable empirical information. Two importantly different cases must now be considered.

[a] H contains no occurrences of any T_j's and is therefore an O-sentence. Thus we can write $H'(\underline{Q})$ for $H(\underline{Q}, \underline{T})$. In this case, (4) is actually equivalent to:

(6) $\vdash [(\exists \underline{t})G(\underline{Q}, \underline{t}) \Rightarrow H'(\underline{Q})]$.

(3) and (6) tells us that $G(\underline{Q}, \underline{T})$ and its Ramsey-sentence $(\exists \underline{t})G(\underline{Q}, \underline{t})$ have the same O-consequences and are *a fortiori* observationally identical. Should $H'(\underline{Q})$ furthermore be a basic statement, then (6) would enable us, through the experimental verification of $\neg H'(\underline{Q})$, to refute $(\exists \underline{t})G(\underline{Q}, \underline{t})$. Referring to the example given in Section 1: both *A* and *A** could be refuted by a basic statement of the form $F(a) \wedge \neg K(a)$. Thus the Ramsey-sentence of every scientific theory has empirical import and does not convey information *only* about the cardinality of its domain of discourse. The question arises as to whether some (non-empirical) reason could make us accept the fully-fledged hypothesis $G(\underline{Q}, \underline{T})$ over and above its logically weaker Ramsey-sentence $(\exists \underline{t})G(\underline{Q}, \underline{t})$. Let us explain why the answer to this question must be negative.

We recall that X* denotes the Ramsey-sentence of an arbitrary proposition X. In his otherwise first-rate 1999, Psillos allowed himself to be misled by Carnap into holding that any structural realist who postulates X* must also accept the *analytic* proposition $X^* \Rightarrow X$ and hence infer X. Therefore, such a realist is anyway compelled to accept the fully-fledged theory X. (Psillos 1999, pp. 51–69). When Psillos furthermore claims that the truth of the *interpreted* hypothesis X is an empirical matter, he presupposes that an interpretation not only of the observational, but also of the theoretical predicates can and must be specified in terms of natural kinds. But this is precisely where the problem lies, for whereas X could in principle be refuted through the falsification of any of its purely observational consequences, *ascertaining* the truth of X must go through the interpretation of its theoretical terms. The latter however refer to the external world; so that, barring what Putnam rightly calls a mystical insight, we are unable to break out of our intentional sphere and fix a semantic relation between our theoretical vocabulary and some *transcendent* reality.

The structural realist in no way denies that scientific theories are physically interpretable; he simply points to the *epistemological* fact that such an interpretation cannot be effectively carried out by us. Though capable of giving a syntactic-structuralist account of 'natural kinds', the structural realist refuses to accept the *material* implication $X^* \Rightarrow X$, which he correctly regards not as an analytic, but as a metaphysical proposition. According to structuralism, there exist no experimental, theoretical, or even aesthetic reason which would make us opt for $G(\underline{Q}, \underline{T})$ in preference to the logically weaker Ramsey-sentence $(\exists \underline{t})G(\underline{Q}, \underline{t})$: for we have seen that $G(\underline{Q}, \underline{T})$ and $(\exists \underline{t})G(\underline{Q}, \underline{t})$ have exactly the same O-consequences; and no unity or simplicity criteria could possibly demarcate between $G(\underline{Q}, \underline{t})$ and $(\exists \underline{t})G(\underline{Q}, \underline{t})$, since, apart from the existential quantifier $(\exists \underline{t})$, the two formulas have the same syntactical structure.

[b] In this second special case of (3), $H(\underline{Q}, \underline{T})$ is assumed to contain no occurrence of any observational predicates Q_i. Thus we can write $H''(\underline{T})$ for $H(\underline{Q}, \underline{T})$. (5) is therefore the statement:

(7) $\vdash [(\exists \underline{t})G(\underline{Q}, \underline{t}) \Rightarrow (\exists \underline{t})H''(\underline{t})]$.

$H''(\underline{T})$ could e.g. be the the sentence $(\exists \underline{q})G(\underline{q}, \underline{T})$, so that, by (6):

(7') $\vdash [(\exists \underline{t})G(\underline{Q}, \underline{t}) \Rightarrow (\exists \underline{t})(\exists \underline{q})G(\underline{q}, \underline{t})]$, which is a logical truth.

This important special subcase will arise if all the terms contained in $G(\underline{Q}, \underline{T})$ are regarded as theory-laden and hence as essentially theoretical. This 'modern' view which, following Brentano, Russell, Poincaré, Grover Maxwell, and Ayer, we regard as deeply mistaken, would mislead us into quantifying over *all* the predicates occurring in $G(\underline{Q}, \underline{T})$, which would then yield, not $(\exists \underline{t})G(\underline{Q}, \underline{t})$, but $(\exists \underline{t})(\exists \underline{q})G(\underline{q}, \underline{t})$ as the Ramsey sentence of $G(\underline{Q}, \underline{T})$ (see above, 1.7). Whatever the case may be, $(\exists \underline{t})H''(\underline{t})$ could well express a tautology, a logical contradiction or a cardinality constraint on the underlying domain of discourse of $G(\underline{Q}, \underline{T})$. This however represents no serious threat to the structuralist position, for it has already been remarked that (5) is but a weak consequence of (4).

Note that we have here Russell's oversight which was seized upon, first by Newman, then by Demopoulos and Friedman. As already mentioned, Russell claimed that "it would seem that whenever we infer from perception, it is only structure that we can validly infer; and structure is what can be expressed by mathematical logic." It looked as though, after we had used some observational results in order to arrive at a structural theory, we could simply jettison these results and be left with a description of pure structure expressible by means of mathematical logic alone. Had Russell stuck to the original Ramsey-sentence, i.e. to the 'mixed' proposition $(\exists \underline{t})G(\underline{Q}, \underline{t})$ instead of $(\exists \underline{t})H''(\underline{t})$, then he need not have given in to Newman's argument.

Let us consider two limiting cases which might arise, should H contain no observational predicates $\underline{Q}$. Taking m = 1 and assuming that $H''(T_1) \equiv (\forall x)(T_1(x) \Leftrightarrow (x \neq x))$, we have a situation in which $(\exists \underline{t}_1)H''(\underline{t}_1)$, i.e. $(\exists t_1)(\forall x)(t_1(x) \Leftrightarrow (x \neq x))$, is a logical truth [for t_1 can be chosen to be the predicate $(\lambda x)(x \neq x)$]. Another example consists of taking $H''(T_1) \equiv [(\forall x)T_1(x) \wedge (\forall x)(T_1(x) \Leftrightarrow (x \neq x))]$; in which case $(\exists \underline{t}_1)H''(\underline{t}_1)$, i.e. $(\exists t_1)[(\forall x)t_1(x) \wedge (\forall x)(t_1(x) \Leftrightarrow (x \neq x))]$, is a logical contradiction; so that our original theory, namely $G(\underline{Q},\underline{T})$, must also be inconsistent [by (1) and (7)].

So far we have conveyed no information about the cardinality of our domain of discourse. Such information will clearly be forthcoming only in the case of finite domains; for if $H''(\underline{T})$ is first-order and has one infinite model, then $H''(\underline{T})$, and hence also $(\exists \underline{t})H''(\underline{t})$ have infinite models of arbitrarily large cardinalities (by the 'upward' Löwenheim-Skolem tjheorem). As for the finite case, it arises if $H''(\underline{T})$ were for example to be a formula of the form: $[(\forall x)T_1(x) \wedge (\exists x_1,x_2,\ldots,x_r)((x_1 \neq x_2) \wedge (x_1 \neq x_3) \wedge \ldots \wedge (x_{r-1} \neq x_r) \wedge (\forall x)(T_1(x) \Leftrightarrow ((x = x_1) \vee (x = x_2) \vee \ldots \vee (x = x_r))))]$, where r is a specific natural number. $(\exists t_1)H''(t_1)$ would then be equivalent to the assertion that the domain of individuals consists of exactly r members.

To repeat, the Ramsey sentence $(\exists \underline{t})G(\underline{Q}, \underline{t})$ reduces to $(\exists \underline{t})H''(\underline{t})$ only if n = 0; i.e. if there are no observational predicates or alternatively if no distinction is made between the observational and the theoretical terms. This is however far from being the position of prominent structural realists like Poincaré, Duhem, or Russell. All the structural realist needs to claim is that our knowledge of the theoretical predicates $T_1, \ldots, T_m$ is encapsulated by the nexus $G(Q_1,\ldots,Q_n,T_1,\ldots,T_m)$; i.e. $T_1, \ldots, T_m$ are known exclusively by description; namely through their mutual relations and through the way—described by $G(Q_1,\ldots, Q_n, T_1,\ldots, T_m)$—in which they link up with the observational predicates $Q_1, \ldots, Q_n$; as for $Q_1, \ldots, Q_n$, they are known *both* by description, *via* $G(Q_1,\ldots,Q_n,T_1,\ldots,T_m)$, and by acquaintance, i.e. phenomenologically or ostensively. Thus the Demopoulos-Friedman objection rests on the unwarranted assumption that the structural realist is bound to deny the observational-theoretical distinction.

[C] Putnam's Model-Theoretic Argument(s)

Prima facie, Putnam's results constitute the most powerful argument against scientific realism—whether classical or structural. Putnam showed that given *any* epistemically ideal theory T_I, we have *no* reason not to regard

T_I *as being true.* Moreover, given certain cardinality constraints, T_I can be proved to possess a 'standard model' in the world, i.e. a model in which all the observational notions have their commonly accepted meanings. This would clearly entail that the Ramsey-sentence T_I^* of T_I *is true simpliciter* (see Putnam 1983b). In order to assess the relevance and scope of Putnam's theorem, it seems to us necessary to describe—if only briefly—his proof together with its hidden lemmas, i.e. together with its covert presuppositions. We shall then examine whether structural realism is really undermined by Putnam's results.

Putnam denotes by S the set of all observable entities. Thus $S \subseteq W$, where 'W' denotes the whole universe. Since the results of all measurements are signified by rational numbers, S can be taken to be enumerable. A set of empirical predicates $\{\pi_i : i \in \Omega\}$ is then specified, where, for each $i \in \Omega$ and for all $\alpha, \beta, \ldots, \gamma$ in S, $\pi_i(\alpha, \beta, \ldots, \gamma)$ is supposed to be ascertainable as true or as false. An epistemically ideal theory T_I is then defined as any consistent first-order system which possesses a set of terms $\{t_\alpha : \alpha \in S\}$ such that $T_I \vdash (t_\alpha \neq t_\beta)$ for any two distinct elements α and β of S. For each $i \in \Omega$, T_I is furthermore required to possess a predicate P_i which has the same number of argument-places as π_i and is such that, for all $\alpha, \beta, \ldots, \gamma$ in S, we have: $T_I \vdash P_i(t_\alpha, t_\beta, \ldots, t_\gamma)$ if $\pi_i(\alpha, \beta, \ldots, \gamma)$ is true, and $T_I \vdash \neg P_i(t_\alpha, t_\beta, \ldots, t_\gamma)$ if $\pi_i(\alpha, \beta, \ldots, \gamma)$ is false.

Let J be any model of T_I. Put:

(8) $\tau_\alpha =_{Def} J(t_\alpha)$ for all $\alpha \in S$.

It can easily be shown that $\pi_i(\alpha, \beta, \ldots, \gamma) \equiv J(P_i)(\tau_\alpha, \tau_\beta, \ldots, \tau_\gamma)$, for all $\alpha, \beta, \ldots, \gamma$ in S. Suppose that $\pi_i(\alpha, \beta, \ldots, \gamma)$ is true. By hypothesis, $T_I \vdash P_i(t_\alpha, t_\beta, \ldots, t_\gamma)$. Since J is a model of T_I, then $J(P_i)(J(t_\alpha), J(t_\beta), \ldots, J(t_\gamma))$, i.e. $J(P_i)(\tau_\alpha, \tau_\beta, \ldots, \tau_\gamma)$, is true. And if $\pi_i(\alpha, \beta, \ldots, \gamma)$ is false, i.e. not-true, then again by hypothesis, $T_I \vdash \neg P_i(t_\alpha, t_\beta, \ldots, t_\gamma)$; hence $J(\neg P_i(t_\alpha, t_\beta, \ldots, t_\gamma))$ is true; that is: $J(P_i(t_\alpha, t_\beta, \ldots, t_\gamma))$, or $J(P_i)(\tau_\alpha, \tau_\beta, \ldots, \tau_\gamma)$, is false. Thus:

(9) $J(P_i)(\tau_\alpha, \tau_\beta, \ldots, \tau_\gamma) \equiv \pi_i(\alpha, \beta, \ldots, \gamma)$, for all $\alpha, \beta, \ldots, \gamma$ in S.

J will be said to be standard if, for all $\alpha \in S$, we have: $J(t_\alpha) = \alpha$, i.e. if $\tau_\alpha = \alpha$. (9) entails that in a standard model:

(10) $J(P_i)(\alpha, \beta, \ldots, \gamma) \equiv \pi_i(\alpha, \beta, \ldots, \gamma)$, for all $\alpha, \beta, \ldots, \gamma$ in S;

i.e., for all i in Ω, $J(P_i)$ coincides with π_i on S.

Although an epistemically ideal hypothesis might be very difficult to construct in practice, it will be assumed, if only for the sake of argument, that such a theory is in principle achievable.

Putnam established that every epistemically ideal T_I has a standard model; for by the Löwenheim-Skolem theorem, T_I has at least one (countable) model, J say. Letting D be the domain of J, we have, by (8): $\{\tau_\alpha\colon \alpha \in S\} = \{J(t_\alpha)\colon \alpha \in S\} \subseteq D$. Thus:

(11) $D = \{\tau_\alpha\colon \alpha \in S\} \cup E$ where $E = D \backslash \{\tau_\alpha\colon \alpha \in S\}$. Hence $|D| = |\{\tau_\alpha\colon \alpha \in S\}| + |E|$.

Since $T_I \vdash t_\alpha \neq t_\beta$ for $\alpha \neq \beta$ and since J is a model of T_I, we have $J(t_\alpha) \neq J(t_\beta)$,

(12) I.e.: $\tau_\alpha \neq \tau_\beta$ for $\alpha \neq \beta$. Therefore $|\{\tau_{\alpha:}\ \alpha \in S\}| = |S|$. By the above:

(13) $|D| = |S| + |E|$. Now define the one-one map f with domain D as follows:

(14) $f(\tau_\alpha) = \alpha$ for all $\alpha \in S$;

and for any $\mu \in E$, define $f(\mu)$ in any way consistent with the requirement that the resulting f be a bijection whose domain is the whole of D.

Now determine the interpretation J' as follows. The domain of J' will be the set:

(15) $D' =_{\text{Def}} S \cup f[E] = \{f(\tau_\alpha)\colon \alpha \in S\} \cup \{f(\mu)\colon \mu \in E\}$

For any function letter (or name) g, let

(16) $J'(g)(\xi, \eta, \ldots, \zeta) =_{\text{Def}} f(J(g)(f^{-1}(\xi), f^{-1}(\eta), \ldots, f^{-1}(\zeta)))$, for any $\xi, \eta, \ldots, \zeta$ in D'.

Finally, for any predicate R, define:

(17) $J'(R)(\xi, \eta, \ldots, \zeta) \equiv_{\text{Def}} J(R)(f^{-1}(\xi), f^{-1}(\eta), \ldots, f^{-1}(\zeta))$, for all $\xi, \eta, \ldots, \zeta$ in D'.

As is well-known, (16) entails:

(18) $J'(u) = f(J(u))$, for all terms u.

Trivially, since J is a model of T_I, then so is J' (for J' can be considered as obtained by relabelling the elements of J). J' is furthermore a standard model, since:

(19) $J'(t_\alpha) = f(J(t_\alpha))$ [by (18)] $=_{\text{Def}} f(\tau_\alpha) =_{\text{Def}} \alpha$ [by (14)].

This result can be strengthened by imposing constraints on the cardinalities of various sets. Let us henceforth assume that the world W is infinite and that T_I has at least one infinite model; then by the 'upward'

Löwenheim-Skolem theorem, T_I has a model having cardinality |W|. In the above, it can therefore be assumed that |D| = |W|. Let us moreover suppose that |W| > |S|. This assumption is not unreasonable, for S is countable while the set of all physical events has at least the power of the continuum. We can thus write:

(20) $W = S \cup S'$, where $S' =_{\text{Def}} W \backslash S$. Hence $|W| = |S| + |S'|$.

Since W is infinite and |W| > |S|, it follows that:

(21) $|W| = |S'|$.

Thus $|W| = |D| = |S| + |E|$ [by (13)]. So, since |W| is infinite and greater than |S|,

(22) $|W| = |E|$; hence $|D| = |W| = |E|$; and so:

(23) $W = S \cup S'$; $D = \{\tau_{\alpha:}\ \alpha \in S\} \cup E$; where $|S'| = |W| = |E|$.

Let Ψ be any one-one map of E onto S'. Going back to (14), we see that f can be defined as follows:

(24) $f(\tau_\alpha) =_{\text{Def}} \alpha$ for all $\alpha \in S$; and for any $\mu \in E$, let $f(\mu) =_{\text{Def}} \Psi(\mu)$.

By (23) and (24), f maps $\{\tau_{\alpha:}\ \alpha \in S\} \cup E = D$ one-one onto $S \cup S' = W$. Thus W = Range of $f =_{\text{Def}} D'$. By the above, J' is a standard model of T_I in the world W; which clearly entails that the Ramsey-sentence T_I^* of T_I is standardly true in W, or *true simpliciter.* The question now is: should the structural realist be worried about these results? We shall show that the answer is decidedly negative.

Putnam actually provides a partial answer to a question posed by Einstein in his *Weltbild;* namely: is a scientific hypothesis which accounts for *all* the facts uniquely determined by the latter? (see Einstein 1934c) Putnam has in effect shown that truth—rather than uniqueness—might well be attainable. More precisely: given the consistency of an epistemically ideal theory T_I, we have no empirical reason not to suppose the Ramsey-sentence T_I^* of T_I to be true. Given further assumptions about the cardinalities of the world W and of the models of T_I, we can even assert the truth of T_I^*. Qua structural realists, we would thus know that the structure described by T_I is realised in the world. A serious problem would arise only if two ideal theories, T_I and H_I say, were such that T_I^* and H_I^* are logically incompatible; but Jane English has proved that this is never the case if—as is tacitly assumed—an *epistemically ideal* theory entails no false observational statement (i.e. no false O-sentence) (English 1980). It is therefore as unexceptionable for a structural realist to claim that the two structures displayed by T_I and by H_I are realised in W as it is for a mathematician to accept that the same set can simultaneously exhibit an algebraic and a toplogical structure. For example, the same set can be the domain both of a Euclidean ring and of a discrete topology.

The more demanding structural realist can go much further than this, but only by appealing to a metaphysical principle similar to the one invoked by both Poincaré and Einstein. The latter put it as follows: the world *is* the realisation of what is mathematically simplest, while Poincaré held the degree of unity-cum-simplicity of a hypothesis to be an index of its truth-likeness (verisimilitude) (see above, 2.2). Despite being incapable of *epistemologically* distinguishing between a theory and its Ramsey-sentence, i.e. between T_I and T_I*, we can certainly adjudicate between T_I* and H_I*. We might thus come to the conclusion that T_I* is more unified than H_I*; in which case only T_I* will be taken to reflect—if only approximately—the real structure of W. In other words, only in the case of T_I* should we realistically interpret the (second-order) quantifications over the theoretical predicates. According to Poincaré, we would otherwise be attributing 'too great a role to chance'; i.e. the excess 'organic compactness' of T_I* over H_I* would have to be regarded as an accidental feature of T_I*; and we know that theoretical physicists of Poincaré's or of Einstein's temper of mind are most reluctant to regard such accidents as ultimately inexplicable. (Both scientists took the degree of unity or of simplicity of a hypothesis to be the expression of one of its *objective* properties; and this in the same sense in which a polynomial of the nth degree is objectively simpler than one of the (n + 1)st degree.)

The above considerations also show that a structural realist can account for the notion of natural kind without abandoning his syntactic-structuralist stance. For the best guide as to what kinds are natural is the degree of unity of an epistemically ideal theory like T_I or—more generally—of any highly successful hypothesis *T* in which such kinds play a central role. In other words, natural kinds are precisely those signified by the predicate and function symbols which give *T* its most coherent formulation. Let us repeat that both the empirical content and what Einstein called the degree of internal perfection of a theory *T* are also possessed by its Ramsey-sentence *T**;* so that there is nothing to choose between *T* and *T**. Note also that a most ardent defender of natural kinds, namely David Lewis, grants that those kinds are natural which are talked about by our best theories (Lewis 1970, p. 428). But what are these 'best' theories? It seems to us that they are precisely those highly unified hypotheses which save the facts without adhoc assumptions. Thus we do not start from a prior description of natural kinds and then base our conjectures on them; but we construct various empirically adequate hypotheses and then select, from among them, the most unified theory, T say. In such a case, T^* expresses exactly what we want to assert; namely that the theoretical predicates which are quantified over in T^* *are really actualised in the world.* In this sense, the Keplerian (near-)ellipses, as opposed to the complex curves which would have to be postulated by a Tychonic reformulation of Kepler's laws, really exist. Similarly: let E be Einstein's gravitational theory and let N denote its 'Euclidean reformulation'; N is known to be empirically equivalent to E, at any rate within a large region of the spatio-temporal manifold. Scientists—who normally are demanding realists—do not look upon the unity displayed by the structure of E, as opposed to the adhoc character of N, as an accidental feature of E. They hold that E describes reality more adequately than does N. Let us finally add that no ideal theory like T_I has so far been put forward; and it seems doubtful that such a theory is at all achievable *in practice*. So science must content itself with provisional hypotheses, or rather with the latter's Ramsey-sentences which are at best *compatible* with all *known* experimental results. It is precisely in these situations that structural realism proves its mettle. For let two such hypotheses—call them T_1 and T_2—be empirically identical. Thus both T_1 and T_2 could subsequently be refuted—by the same facts. Yet T_2 —but not T_1—might possess a structure which points, i.e. which heuristically leads to a more powerful theory, S say; where S empirically supersedes both T_1 and T_2. For example, S might explain some of the results which undermined T_1 and hence also T_2; yet, by virtue of the Correspondence Principle, only the structure of T_2 survives—possibly in a modified form—in S. Here, we can again take, as example, the case where: $S \equiv$ Quantum Mechanics; $T_1 \equiv$ Newtonian Dynamics; and $T_2 \equiv$ (Classical) Hamiltonian Dynamics (see above, Chapters 2 and 4). This example is instructive, first because it does not rely on hindsight: long before the birth of Quantum Physics, Poincaré had noted that despite their strict observational equivalence, T_1 and T_2 have very different structures and hence make different statements about the world (see above, 1.1). Secondly: as we go over from T_2 to S, only the mathematical structure, as typified by the form of the Hamiltonian function, is preserved; even the logical types of this function's arguments are altered: whereas the arguments of the classical Hamiltonian are functions of the time-variable, those of the quantum-mechanical Hamiltonian are operators, i.e. higher-order entities, namely functions of functions. No example illustrates as clearly as the present one the essential role played by the notion of structure in the development of fundamental science.

It might be maintained that structural *realism* does not deserve its name. This of course is a purely terminological objection. We can nonetheless point out that the most prominent advocates of structural realism, namely Duhem and Poincaré, were realists in the intuitive sense of believing in the existence of a structured and mind-independent reality which the scientist tries to capture. Moreover, according to them as well as to modern realists like MacMullin, we are entitled to regard some of our concepts as *metaphors,* which nevertheless model the external world; but in this case, any question of a strict *term-by-term* reference to anything at all, whether it be intended or not, becomes otiose (MacMullin 1984).

Bibliography

Albert, H. 1975. *Traktat über kritische Vernunft*. Tübingen: Mohr (Siebeck).

Ayer, A.J. 1956. *The Problem of Knowledge*. London: Penguin.

———., ed. 1959. *Logical Positivism*. USA: The Free Press.

Barrow, J.D., and Tipler, F.J. 1986. *The Anthropic Cosmological Principle*. Oxford: Oxford University Press.

Bell, E.T. 1937. *Men of Mathematics*. New York: Simon and Schuster.

Bell, J.L. 1977. *Boolean-valued Models and Independence Proofs in Set Theory*. Oxford: Oxford University Press.

Benacerraf, P., and H. Putnam, eds. 1983. *Philosophy of Mathematics*. Cambridge: Cambridge University Press.

Bernays, P. 1976a. *Abhandlungen zur Philosophie der Mathematik*. Darmstadt: Wissenschaftliche Buchgesellschaft.

———. 1976b. Probleme der theoretischen Logik. In Bernays 1976a.

Bouveresse, R. 1998. *Karl Popper ou le rationalisme critique*. Paris: Vrin.

Boyer, A. 1992. Physique de croyant? Duhem et l'autonomie de la science. *Revue internationale de philosophie* 92, Vol.46, No. 182.

Brentano, F. 1924. *Psychologie vom empirischen Standpunkt (Band 1)*. Hamburg: Felix Meiner.

———. 1930. *Wahrheit und Evidenz*. Hamburg: Felix Meiner.

———. 1968. *Psychologie vom empirischen Standpunkt (Band 3)*. Hamburg: Felix Meiner.

———. 1982. *Deskriptive Psychologie*. Hamburg: Felix Meiner.

Carnap, R. 1966. *Philosophical Foundations of Physics*. New York: Basic Books.

Chihara, P. 1990. *Constructibility and Mathematical Existence*. Oxford: Oxford University Press.

Demopoulos, D., and Friedman, M. 1985. Critical Notice: Bertrand Russell's *The Analysis of Matter*: Its Historical Context and Contemporary Interest. *Philosophy of Science* 52; pp. 621–663.

Descartes, R. 1986. *Meditations on First Philosophy*. Trans. J. Cottingham. Cambridge: Cambridge University Press.

Dugac, P. 1976. *Richard Dedekind*. Paris: Vrin.

Duhem, P. 1914. *La théorie physique: son objet, sa structure*. Paris: Marcel Rivière.

Eddington, A.S. 1923. *The Mathematical Theory of Relativity*. Cambridge: Cambridge University Press.

———. 1958. *The Philosophy of Physical Science*. Ann Arbor: Unversity of Michigan Press.

Einstein, A. 1923. The Foundations of the General Theory of Relativity. In Einstein et al. 1923.

———. 1934a. *Mein Weltbild.* Frankfurt am Main: Ullstein.

———. 1934b. Geometrie und Erfahrung. In Einstein 1934a.

———. 1934c. Prinzipien der Forschung. In Einstein 1934a.

———. 1949. Autobiographocal Notes. In Schilpp 1949.

———. 1994. *Ideas and Opinions.* New York: Modern Library.

Einstein, A., et al. 1923. *The Principle of Relativity.* New York: Dover.

English, J. 1980. Underdetermination: Craig and Ramsey. *The Journal of Philosophy,* pp.453–462.

Feyerabend, P.K. 1975. *Against Method.* London: Humanities Press.

Field, H. 1980. *Science Without Numbers.* Oxford: Blackwell.

Fine, A. 1986. *The Shaky Game.* Chicago: Chicago University Press.

Frege, G. 1986. *Funktion, Begriff, Bedeutung.* Göttingen: Vandenhoeck und Ruprecht.

Gödel, K. 1983a. What is Cantor's Continuum Problem? In Benacerraf and Putnam 1983, pp. 470–485.

———. 1983b. Russell's Mathematical Logic. In Benacerraf and Putnam 1983, pp. 447–469.

Grünbaum, A. 1973. *Philosophical Problems of Space and Time.* Dordrecht: Reidel.

Heidegger, M. 1953. *Einführung in die Metaphysik.* Tübingen: Niemeyer.

Heinzmann, G. 1985. *Entre Intuition et Analyse.* Paris: Albert Blanchard.

Hermes, H. 1955. *Einführung in die Verbandstheorie.* Berlin: Springer.

Husserl, E. 1928. *Ideen (I).* Tübingen: Niemeyer.

———. 1929. *Formale und transzendentale Logik.* Tübingen: Niemeyer.

———. 1939. *Erfahrung und Urteil.* Hamburg: Felix Meiner.

———. 1969a. *Die Krisis der europäischen Wissenschaften und die transzendentale Phänomenologie.* Hamburg: Felix Meiner.

———. 1969b. *Cartesianische Meditationen.* Hamburg: Felix Meiner.

Jacobson, N. 1953. *Lectures in Abstract Algebra (Vol. 2).* New York: Van Nostrand.

Kant, I. 1787. *Kritik der reinen Vernunft.* Hamburg: Felix Meiner.

———. 1953. *Prolegomena to any Future Metaphysics.* Manchester: Manchester University Press.

Kilmister, C.W. 1984. *Russell.* Brighton: Harvester Press.

Kraft, V. 1968. *Der Wiener Kreis.* Wien/New York: Springer Verlag.

Kretschmann, K. 1917. Über den physikalischen Sinn der Relativitätspostulate, A. Einsteins neue und seine ursprüngliche Relativitätstheorie. *Annalen der Physik* 53.

Lakatos, I. 1976. *Proofs and Refutations.* Cambridge: Cambridge University Press.

———. 1978. *The Methodology of Scientific Research Programmes.* Cambridge: Cambridge University Press.

Lakatos, I., and E. Zahar. 1978. Why Did Copernicus's Programme Supersede Ptolemy's? In Lakatos 1978.

Leplin, J., ed. 1984. *Scientific Realism.* Berkeley: University of California Press.

Lewis, D. 1970. How to Define Theoretical Terms. *The Journal of Philosophy,* Vol. 57, No. 13; pp. 427–446.

Lorentz, H.A. 1895. Versuch einer Theorie der elektrischen und optischen Erscheinungen in bewegten Körpern. Reprinted in Lorentz 1934, Vol. 5.

———. 1904. Electromagnetic Phenomena in a System Moving with any Velocity Less than that of Light. Reprinted in Lorentz 1934, Vol. 5.

———. 1909. *The Theory of Electrons.* New York: Dover.

———. 1934. *Collected Papers.* The Hague: Nijhoff.

Mach, E. 1922. *Die Anlyse der Empfindungen.* Darmstadt: Wissenschaftliche Buchgesellschaft.

Mach, M. 1926. *Erkenntnis und Irrtum.* Leipzig: J.A. Barth.

Mach, E. 1933. *Die Mechanik.* Darmstadt: Wissenschaftliche Buchgesellschaft.

MacMullin, E. 1984. A Case for Scientific Realism. In Leplin 1984, pp. 30–36.

Malament, D. 1982. Review of Field 1980. *Journal of Philosophy* No. 79, pp. 523–534.

(Grover) Maxwell.1971. Structural Realism and the Meaning of Theoretical Terms. In M. Radner and S. Minokur, eds., 1971. *Minnesota Studies in the Philosophy of Science*, Vol.8, pp. 181–192.

McGuinness, B.F. 1967. *Friedrich Waismann: Wittgenstein und der Wiener Kreis.* Oxford: Blackwell.

Meyerson, E. 1907. *Identité et réalité.* Paris: Payot.

Meyerson, E. 1921. *De l'explication dans les sciences.* Paris: Payot.

Newman, M.H.A. 1928. Mr. Russell's 'Causal Theory of Perception'. *Mind* Vol. 37, No. 146, pp. 137–148.

Papineau, D., ed. 1996. *The Philosophy of Science.* Oxford: Oxford University Press.

Paty, M. 1988. *La matière dérobée.* Paris: Archives contemporaines.

Planck, M. 1965a. *Vorträge und Erinnerungen.* Darmstadt: Wissenschaftliche Buchgesellschaft.

———. 1965b. Die Einheit des physikalischen Weltbildes. In Planck 1965a.

———. 1965c. Sinn und Grenzen der exakten Wissenschaft. In Planck 1965a.

———. 1965d. Kausalgesetz und Willensfreiheit. In Planck 1965a.

Poincaré, H. 1881. Sur les fonctions fuchsiennes. *Comptes rendus de l'Académie des Sciences.*

———. 1895. À propos de la théorie de M. Larmor. Paris: *Éclairage électrique.*

———. 1900. La théorie de Lorentz et le principe de réaction. *Arch. Neerland. Sci. Exact.* 5.

———. 1901. *Électricité et optique.* Paris: Gauthier-Villars.

———. 1902. *La science et l'hypothèse.* Paris: Flammarion. All page references are to the English translation (Poincaré 1952).

———. 1906a. *La valeur de la science.* Paris: Flammarion. All page references are to the English translation.

———. 1906b. Sur la dynamique de l'électron. *Rendiconti del Circolo Matematico di Palermo.*

———. 1908. *Science et méthode.* Paris: Flammarion. All page references are to the English translation (Poincaré 1958).

———. 1913a: *Dernières Pensées.* Paris: Flammarion. All page references are to the English translation (Poincaré 1963).

———. 1913b. La Logique de l'Infini. In Poincaré 1913a, p. 95.

———. 1916. *Œuvres.* Paris: Gauthier-Villars.

———. 1952. *Science and Hypothesis.* New York: Dover.

———. 1958a. *Science and Method.* New York: Dover.

———. 1958b. *The Value of Science.* New York: Dover.

———. 1963. *Last Essays.* New York: Dover.

Popper, K.R. 1959. *The Logic of Scientific Discovery.* London: Hutchinson.

———. 1963. *Conjectures and Refutations.* London: Routledge.

———.1972. *Objective Knowledge.* Oxford: Oxford University Press.

———. 1979. *Die beiden Grundprobleme der Erkenntnistheorie.* Tübingen: Mohr (Siebeck).

———. 1982. *The Open Universe.* London: Hutchinson.

Psillos, S. 1999. *Scientific Realism.* London: Routledge.

Putnam, H. 1983a. *Realism and Reason.* Cambridge: Cambridge University Press.

———. 1983b. Models and Reality. In Putnam 1983a.

Quine, W.v.O. 1980a. *From a Logical Point of View.* Cambridge, Ma: Harvard University Press.

———. 1980b. On What There Is. In Quine 1980a.

———. 1980c. Two Dogmas of Empiricism. In Quine 1980a.

Redhead, M. 1975. Symmetry in Intertheory Relations. *Synthese* 32. Dordrecht: Reidel.

———. 1987. *Incompleteness, Nonlocality, and Realism*, Oxford: Oxford University Press.

———. 1995. *From Physics to Metaphysics.* Cambridge: Cambridge University Press.

Reichenbach, H. 1928. *Philosophie der Raum-Zeit-Lehre.* Berlin: Walter de Gruyter.

Russsell, B. 1927. *The Analysis of Matter.* London: Allen and Unwin.

———. 1973. *Essays in Analysis.* London: Allen and Unwin.

Schilpp, P.A., ed. 1949. *Einstein: Philosopher-Scientist.* New York: Tudor.

Schlick, M. 1920. *Space and Time.* New York: Dover.

———. 1938. *Gesammelte Aufsätze.* Wien: Springer.

———. 1986. *Philosophische Logik.* Frankfurt am Main: Suhrkamp.

Schopenhauer, A. 1958. *The World as Will and Representation.* New York: Dover.

Schrödinger, E. 1950. *Space-Time Structure.* Cambridge: Cambridge University Press.

Stillwell, J., ed. (and translator). 1985. *Henri Poincaré: Papers on Fuchsian Functions.* New York: Springer.

Tragesser, R.S. 1984. *Husserl and Realism in Logic and Mathematics.* Cambridge: Cambridge University Press.

Watkins, J.W.N. 1984. *Science and Scepticism.* Princeton: Princeton University Press.

Whitehead, A.N. 1926. *Science and the Modern World.* Cambridge: Cambridge University Press.

Whittaker, E.T. 1953. *A History of the Theories of Aether and Electricity: the Classical Theories.* London: Thomas Nelson.

Wittgenstein, L. 1972. *Tractatus Logico-Philosophicus.* London: Routledge.

Worrall, J. 1996. Structural Realism. In Papineau 1996.

Zahar, E. 1977. Mach, Einstein, and the Rise of Modern Science. *British Journal for the Philosophy of Science* 28.

———. 1989. *Einstein's Revolution: A Study in Heuristic.* La Salle, Illinois: Open Court.

Index